Fireproof
Homebuilding

Fireproof Homebuilding

Leo Du Lac

McGraw-Hill, Inc.

New York   San Francisco   Washington, D.C.   Auckland   Bogotá
Caracas   Lisbon   London   Madrid   Mexico City   Milan
Montreal   New Delhi   San Juan   Singapore
Sydney   Tokyo   Toronto

©1995 by **Leo Du Lac.**
Published by McGraw-Hill, Inc.

hc   1 2 3 4 5 6 7 8 9 0 DOC/DOC 9 9 8 7 6 5

Library of Congress Cataloging-in-Publication Data
Du Lac, Leo.
   Fireproof homebuilding / by Leo Du Lac.
     p. cm.
   Includes index.
   ISBN 0-07-018091-1
   1. Building, Fireproof. I. Title.
TH1065.D85 1995
693'.82—dc20   95-56
                                      CIP

Acquisitions editor: April D. Nolan
Executive editor: Robert E. Ostrander
Production team: Katherine G. Brown, Director
                   Ollie Harmon, Coding
                   Janice Ridenour, Computer Artist
                   Toya B. Warner, Computer Artist
                   Wanda S. Ditch, Desktop Operator
                   Jodi L. Tyler, Indexer
Design team: Jaclyn J. Boone, Designer
               Katherine Stefanski, Associate Designer

WK2
0180911

Contents

Foreword

Leo Du Lac became aware of a greater need for fire safety in the United States a number of years ago, while he was involved in the restoration of fire-damaged buildings. He then discovered that the number of people killed in fires in the United States has been consistently the highest among first-world nations. With more intensive fire-prevention campaigns by fire departments, this figure has dropped. But thousands of tragic deaths still occur annually.

We've all heard the stories. A fireman dropped in at a kindergarten one afternoon to acquaint the youngsters with fire prevention. One boy told of what had happened at his house. His father lit a fire in the fireplace, and then father and son went off to the grocery store. When they returned, their house was on fire. Obviously, for a 5-year-old little boy, the experience was frightening and traumatic. And he was lucky: He survived.

Then there was the couple who came home from a party a little drunk. After they were in bed, a fire started in the kitchen where a bit of finish plaster had flaked off from water damage. The fire eventually consumed all the oxygen in the bedroom, which was far enough from the kitchen that it barely even got warm from the heat of the fire. The couple smothered from poisonous gases, the product

xi

of modern fixtures and lack of oxygen. If their bedroom door had been closed and a window had been left partially open, or if they had repaired the damaged plaster that caused the fire in the first place, they could have survived the fumes. A smoke detector in the kitchen might have provided an early warning.

But even smoke detectors can't solve the problem of so many residential fires. Certainly, the warning that smoke detectors provide has prevented a great deal of serious injury and death, but, in many cases, the house is still a total loss, consumed by flames or seriously damaged by the water used to douse the fire. Fire insurance doesn't always cover the bill to repair or replace the house, and nothing can replace the photos, memorabilia, and personal loss incurred by such a tragedy.

Obviously, then, smoke detectors alone are not the answer to the problem. What about other forms of prevention? Fire departments are always working to prevent fires within their districts. In many places, firefighters conduct an annual campaign urging people to become better housekeepers. They stress important preventive measures and bad habits that should be avoided. The following are just a few:

➤ Clean the weeds away from all buildings. (In the desert, the regulation is 15 feet, but 30 feet is better.)

➤ Move lumber or other flammable materials away from the house.

➤ Don't store flammable paint or other combustibles in the house.

➤ Never smoke in bed.

➤ Never light a fire in a fireplace when you know you will be leaving the house (and the fire) unattended.

➤ Keep chimneys and fireplaces clean and clear of debris.

➤ Close bedroom doors at night, and leave a window partially open.

➤ Avoid exposed wires and overloaded outlets.

These measures are helpful, but thousands of homes every year burn, nonetheless. Why? Even conscientious people can be careless on occasion, and one occasion is all it takes. The greater problem behind the prevalence of fires in the United States hinges on the fact that American homes, in general, are built of flammable materials and not designed to withstand fires (or many other kinds of natural disasters, for that matter).

Restoring fire-damaged homes is big business in the United States. It won't be that way with new homes when contractors learn to build fireproof as Du Lac recommends. During Du Lac's experience with this restoration, he began to realize that fire prevention really begins with the structure—one built with incombustible materials.

Recently, the United States government sponsored a project to promote sprinkling of new homes as a means of fire prevention. What they failed to recognize is that, for the price of a sprinkling system, one could build a house that would withstand a fire far better than one that had been periodically doused by a sprinkler.

You can search the libraries for material on the subject of an all-fireproof residence. You will probably find little on existing fireproof homes in the United States, except for the occasional house built by an engineer. I know of one engineer who used heavy steel beams for clear span of the rooms in his house; you could have driven a train over the structure. Of course, the cost of such a house would have been prohibitive for the average homeowner, so his methods, understandably, never caught on.

But residential contractors can learn from commercial construction. For example, high-rise apartments are virtually fireproof because the building code requires them to be. The required framing materials (steel, in many cases) lend themselves to a sturdiness that will support concrete roofs and floors, and all of these materials are more durable than even plus-code lumber.

There is no reason the American home cannot be built with concrete roofs and floors. In fact, that is the unique feature of residential construction in Europe, where everything is built fireproof. In the United States, many factories are built of concrete slabs, tilt-up, or an alternate method of concrete construction. If we were to take this one step farther, as they do in the Middle East, the whole structure would be fireproof. So why not apply these commercial construction methods and European traditions to residential construction?

Foreword

The firefighters in Europe are confronted with a situation similar to the Maytag repair man: They become bored from sitting around. The rate of fire damage in Europe, where fireproof construction is prevalent, is 33 cents per person per year. In the United States and Canada, where wood products are readily available, the rate is $4.00 per person per year. This should be sufficient proof that incombustible materials used in the process of building pays off.

This book is a compilation of knowledgeable information that can be used in the design and construction of a fireproof residence. The book does not, however, deal with engineering, which in many places is required for this type of work, giving calculations for each design, proving that the building will withstand the various shocks of nature, wind resistance during a cyclone, and the shifting of the earth's crust. But it will supply information related to upgrading of existing roofs, remodeling buildings, rendering them more fireproof, and general construction tips on how to build the fireproof house.

Leo Du Lac's personal background and involvement in fire-damage restoration make him one of the few people fully qualified to have written this book. He has been in the construction business all his life, with four years devoted to repairing fire-damaged buildings. He has published numerous articles on the subject, and he has studied homebuilding in Italy, France, Spain, Ireland, Sweden, and Germany, where fireproof construction is standard procedure.

Read the book to see where it might apply to building a more fireproof and safer building for your clients. Chances are, whether you are an architect, a contractor, or a homebuilder, you will probably be inspired to improve the fire safety in your own home.

Richard Lupton
Deputy Fire Marshal
Hesperia, California, Fire Department

Introduction

Why don't we, in the United States, build fireproof? If I am forced to answer my own question, I would have to say that we didn't know there was such a thing as a fireproof house. Architects and engineers know how to design a fireproof house, but the public, their clients, is not aware that there is a better way to build, the fireproof method. So architects have few requests for such designs, and they avoid the hassle of explaining fireproof materials to builders and homebuyers.

Fireproof construction is a long-neglected phase of the homebuilding industry in the United States and Canada, primarily because wood is so readily available in North America. Until recently, it was also the most inexpensive building material around.

But, as lumber-price increases of the past few years show, the incentives are increasing for the use of other, more durable materials. As contractors turn away from lumber, they can just as easily turn to residential steel, concrete homebuilding systems, stone, brick, and other incombustible materials.

Across the Atlantic, lumber is not nearly so prevalent, so alternative materials have always been necessary. As a result, the incidence of fire-related property damage, injury, and death is far lower in Europe than in the U.S.

You'll be hard-pressed to find wholly fireproof dwellings being built in any of the tracts in the States. But there is a start: Most commercial buildings are of cement block, especially the firewalls, while some architects are designing the inner partitions of steel studs. Industrial buildings are generally of cement and concrete block. (I am referring primarily to California. Trends will vary in other locales, some favoring steel studs, although I've found only an occasional small residential tract where the contractor has experimented with steel studs.)

Before starting to write this book, I ran around looking for solar-heated and wind-powered homes. I discovered that most contractors are too busy trying to meet tight deadlines to introduce any new concepts that might slow down their work. In the sunbelt, some cities and counties require a solar hot-water heater. This is a start in updating our housing industry, but it's also a roadblock until a practical concept has been implemented to undertake these innovative changes with efficiency. The same needs to be done to find a practical, cost-effective method in the United States for building fireproof homes.

You will find brick houses in the East and brick-and-block houses in the North and Midwest, and even a few stone houses scattered throughout the country. In Florida, concrete block is used extensively for exterior walls because reinforced block stands up better against hurricanes than lumber (and because termites find wood so toothsome). But that is as far as fireproof structures go. The rest of the building is standard construction, which means wood, a combustible product.

At first, contractors will have to charge more to build fireproof homes because they are unfamiliar with the procedure, and, even though some parts of construction might be familiar, others will present a learning challenge for field teams at first. But the materials themselves are no more expensive, particularly in light of the current exhorbitant price of lumber.

The Brick Institute of America experimented with a brick house in the Carolinas and found it to be no more expensive than the frame houses being built. The experimental house sold before it was finished. There were requests for more houses of the same design, while frame houses remained unsold at the time. All it would take to make a brick house fireproof would be to eliminate the lumber used in interior laths and the roof. When contractors catch on to building metal trusses or concrete beams instead of using wood, prices will become competitive.

If you have any thought of building, you should consider the information in this book so that you will never have to rebuild. Even if you are just remodeling, you could upgrade certain phases of your present residence or those of your customers to make them less vulnerable to fire.

1

Starting out right

MY father built the house where I was born, and it was far from the nearest road. The neighbors all considered his residence out of place. Most passersby never knew there was a house in that pleasant valley by a running stream. The joke went around that Joe bought a load of lumber and hauled it cross-country to his farm nine miles from town. Before he got to the road that fronted his property, the neighbors said, his horses encountered a skunk and started to run. In their excitement, the team made a sharp turn, tipped the wagon over, and dumped the lumber, and that was where my father built his house, a half mile from the road. Homebuilders face serious consequences today if they attempt to site a house in such a fashion.

It used to be that in preparing to build a house, a homebuilder bought a load of lumber, hauled it to the site, and started to hammer nails. No thought was given to permits, terrain, floods, or, most importantly, the potential of unforeseen fire hazards. Today, homebuilders have to worry about all these things and more. Each lot has its own potential problems: Floods, earthquakes, fires, possible encroachment, and a bevy of code requirements are just a few of the things homebuilders must consider and plan for with each project. It does little good, then, in a book about fireproof homebuilding, to launch into construction without first ensuring that the lot in question is going to hold the house.

Although most of the following examples of what to do and what not to do prior to building come from the California desert region, you will no doubt find similar adverse situations everywhere. Always look into every eventuality before buying haphazardly and building where you might get wiped out by unforeseen fires or floods.

Fire hazards

If your intended building lot already has streets, gutters, and sewers, you have little to worry about in the way of fire hazards. You need only maintain the required firebreak next to all buildings, which is 15 feet (4.57 m)—30 feet (9.14 m) is even better in the drylands.

2

One look at a lot such as the one shown in Fig. 1-1 will tell you that all native vegetation needs to be scraped off the surface and piled up with a little dirt covering it over as you grade. (The cost of hauling such trash to the dump could be prohibitive.) Code also allows you to destroy the weeds for another 15 feet (4.57 m) on the lot adjoining your vacant land.

Figure **1-1**

The main preventive measures to fireproof a lot include removing the threatening vegetation and grading properly.

These measures should take care of the fire potential for that season. The homeowners will have to maintain the land on an annual basis, raking up native grass if it rains that year. I failed to do this one year, and my front yard burned, destroying two Joshua trees. As some of the native bushes grow back, it's a good idea to destroy some of them, leaving the property with native landscaping that will require no watering.

Flood hazards

When buying a lot, the potential of flooding might be more important than the unforeseen fire hazards until the house is built. The chance of fire damage can be eliminated by grading and by destroying the weeds on a regular basis. But to eliminate flooding problems, you have to buy the right lot. Even then, it's easy to have guessed wrong. I built on the high ground on the back of my lot (Fig. 1-2). Unfortunately, that did not keep my mailbox from being washed away or my driveway from being washed out.

3

Figure 1-2

This innocent-looking location floods after an inch of rain. If there is a flood-control easement on the property where you propose building, position the house on higher ground where it won't be wiped out during a once-in-a-hundred-years deluge.

The possibility of flooding is an even more important consideration in areas where fire is also a danger. The earth's most precious commodity often can cause the worst damage, especially after an extensive fire has destroyed all brush or ground cover.

In my windy city, many homes were built at a time when planning was not much in evidence. During a recent rainy spell, after a fire had destroyed the little available vegetation holding the soil together, flood water paid a visit to an unsuspecting homeowner. Mud and rock pushed his front door open and, in a short time, was escaping out the back door like an army in retreat.

If a house is built where water flows, a flash flood can wash away the house. To make matters worse, in some areas there is little evidence that water has passed over that parcel of land in the past 100 years. Your county has natural drainage maps for most of the land on the market: Do not build on that part of a lot designated as drainage. You might think you can get away with it, that the inspectors might not notice a foundation dug on a drainage easement. They might pass that phase of your job, but keep in mind that, as the homebuilder, you will be held responsible for ensuing damage, not the inspector.

People are moving farther out of the city, in advance of planning and facilities, where unforeseen disasters lie hidden. The pleasant San Fernando Valley north of Los Angeles built up rapidly after a dam in the mountains was built to contain flood waters. The valley was deemed safe for a vast building project.

The usual winter rains came and filled the dam. Late in the spring, with the dry season of the year approaching, no water was drained from the dam. Then came an unexpected wet March; rain poured over the dam in torrents, flowed down the Los Angeles River, and flooded the San Fernando Valley. When all was over, parquet blocks of the newly installed floors swelled up and blocked all the door passages. Since then, a check dam has been built as a further protection for residents of the valley.

By learning from the experiences mentioned here, you, the homebuilder, can avoid a few mistakes in buying a lot in an outlying district where planning is not in effect.

Site planning

As a responsible homebuilder, if you can't find the survey stakes or the owner can't verify them, it's best to have the lot surveyed. Then, when building, follow the setback rules in effect for that city. A clause in the sales contract should state that the former owner is responsible for accurate dimensions, or, with a deposit in escrow, ask that the property be surveyed.

Just recently my neighbors dug the foundation for a garage three feet (0.9 m) from the property line. When the inspector came out, he informed the contractor that the setback for residential in that area of town was 15 feet (4.57 m). The contractor started over, filled in most of the trenches, and re-dug the foundation.

The location of any proposed structure should be on a plot plan, verified by the owner so that no mistake can be made. There is something wrong when a contractor doesn't know the setback regulations in his own town, as happened in this instance.

Drainage

If you are in a rugged area where you are responsible for streets, gutters, and pavement, you might have to have the plot engineered to steer the water down the street. In some desert areas, engineering has to show that rain will be retained on the lot, so precipitation from flash floods will not be dumped onto the neighbor's land.

You are responsible for all water dropping onto your property. In this way, what little water becomes available in the desert usually soaks into the water table to be pumped out, instead of passing on to the next town. The lot must be graded accordingly before building, or in some cases after the building is completed. The owner can't move into the house or rent the property until the homebuilder ensures compliance with these regulations.

There might be exceptions to these requirements, if part of your lot is a natural drainage. In my present location, the water from my neighbor to the west drains onto my property, but I have a berm that holds the water back, and it floods my trees a good deal, even though annual rainfall is but four inches (10.16 cm) a year.

There are always drawbacks to new regulations enacted by the city council or the building department in any region. The homebuilder will rightfully complain that retaining all water on the property for a proposed new residence sometimes requires three times the grading hours that would otherwise have been spent on the land.

In one area, the city gave permission for a small number of homes to be built. When streets were graded and paved, rainfall began to dump onto land below where no problem had ever been encountered before. A home or two was flooded. In this instance, one of the flooded homeowners successfully sued the city. Several other homeowners are now trying to sue the city for water damages. The city had not yet incorporated when some of these owners chose to build in those locations. Now the city is doing more of the thinking for people who have had little experience in locating a building on a lot.

If the lot is large in a dry area, the grading contractor might try to get away with leveling a pad for fruit trees in a retention basin or a berm in the path of the flood-control easement. After each rain, the garden or orchard will flood, and the homeowner won't have to water for a month or six weeks.

But retention basin limits in dry areas are strict, and any good contractor should know that he is not allowed to build on designated flood-control easements. An engineer will show all these factors on a preliminary plan that the city should recognize before allowing a contractor to build.

A four-inch (10.16 cm) annual rainfall can cause so much havoc in the desert because there is little vegetation to hold the soil together (Fig. 1-3).

You must learn to build with caution in any terrain. In hilly country (for example, in eastern Kansas where the glaciers tore up the land before regulations were enacted to stop the damage), there is plenty of rainfall. In any area like this, you might want a retention dam 10 or 20 feet (3.05 or 6.1 m) high in front of the house. With the annual rains, a lake will form, and you could stock it with channel cats. Then, looking down from above, this makes an ideal building site.

Figure 1-3

This canyon was washed out along one of Hesperia, California's, roads in the high desert. There is no vegetation to hold the soil together. If this series of holes were filled with trash under supervision and compaction, the road could again be paved. Houses in situations not even as bad as this were vacated and turned back to the bank.

Utilities

As a contractor, if you are preparing to build a commercial building for lease or rent, don't invest too much money until all utilities are taken into consideration. When it comes to industrial, some cities charge $2,500 for a simple site inspection.

If you are building on speculation, pay careful consideration to land where you are required to foot the bill for all utilities, as well as streets, curbs, and sewers. You might not want to invest the additional monies, but there is always a buyer who can afford the additional investment.

In the case of commercial building, you, the contractor, should obtain three bids on each item, roads, water mains, electrical, and sewers. The bank will want all these figures before they even talk about a loan. The city will place a judgment on such property and will not allow the owner to move into a house or rent the building until these regulations have met compliance.

It's not likely that you will find sewers in new areas where large lots are available. No newly incorporated city will be able to bear the costs, and these are sometimes deferred for 10 or even 20 years. In the meantime, you will have to install septic tanks with 20-foot (6.1 m) seep tanks lined with block. This will probably cost $1,500 or more.

Speculation

During good economic times, a builder who had a loan that covered construction costs (including streets) built an apartment building. For the builder to recover his investment, he had to lease for 60 cents a foot (0.3048 m). Then came a slump, a recession. The building couldn't be rented for 30 cents a foot (0.3048 m), and the owner finally had to relinquish his claim to the bank.

Homebuilders, when not busy, are often speculatively buying property for the next house. Always keep in mind that one of the definitions of the word *speculation* is a weighing of the risks—in this case, of whether or not the ideal lot might have expensive problems. For example, you might think it would be a wonderful selling point to build a home with "the ocean as a front yard." Remember, though, the ocean is restless and always claiming more land every day. It might soon want to include your lot or the house you have just finished building. Retaining walls are of little effect when the waves pound and undermine everything built to protect the property.

Borrowing money

Speculation, of course, often includes borrowing money. At one time I was building a small tract of industrial buildings. The vice-president of the local bank was on the job one morning and liked what he saw. He suggested the bank loan me sufficient money to build up the property, all at one time. I had planned to build one building at a time, each after I had acquired the funds to build and the tenants to buy or lease. If I did what the vice-president suggested, I reasoned, I would be giving the banking institution at least one factory outright, through interest payments. Because I was happy building one building at a time, doing most of the work myself, I chose the safer method.

Borrowing money might depend on the size of the contracting institution doing business and the amount of risk one is willing (or able) to take. The president of a large, prestigious contracting firm would rather drop dead than push a shovel into rocky soil. And there's no incentive to do so when he can afford to spend a little extra capital on perfect building lots. On the other hand, a small contractor, if he is willing to do a little extra work, can get rich with a bank loan and several nicely placed homes on the same rocky soil. If something goes wrong, of course, he can also lose everything.

European &
commercial traditions

IF you study fireproof homes in Europe, you will find that the people who build them are no more fire-conscious than the people in the United States. Europeans build fireproof as a matter of course, partly because of the lack of lumber. They really don't realize they are building fireproof homes, but their rate of annual fire damage per person is far below that in the United States and Canada.

But no one needs to go to Europe to study fireproof construction. One look at the commercial high-rise buildings in the United States, prevalent in the more densely populated parts of the country, will provide a wealth of information on fireproof construction. Fireproof construction is required by code in such areas, and we can apply many of the same methods to residential construction.

A number of architects and engineers from France were here in the high desert recently, studying our methods of home construction. The contractors who showed these men around made them feel like idiots. In France, the Americans told the Europeans they were wasting their time with concrete and noncombustible materials, and they were involved in a difficult and lengthy building process. The contracting firm here knew little about the methods prevalent in France. But I say the French methods of building fireproof are far superior to our habits of wood, wood, wood. We could discontinue many Fire Departments if everything were built of incombustible materials as in Europe.

The European way

While in France recently, to the chagrin of my wife, I was talking to the workmen on a construction job. The two-story house they were working on clearly would never burn (Fig. 2-1). The owners were not aware that the house was being built fireproof, but it included even fireproof floors, as well as a concrete roof covered with tile.

In the United States, most dwellings are built of wood so they last only two or three generations. Then, those that haven't been consumed by a fire are usually added to our already-impressive dumps.

Figure 2-1

In France, they don't designate a structure as fireproof; they build of noncombustible materials out of habit. Homes in Europe are every bit as elegant as anything found in the United States, but they are never eaten up by termites, and they do not burn down.

The house in France where my great-great-grandfather lived was built over 200 years ago and is still standing and inhabited today (Fig. 2-2). It has recently been completely renovated inside. The outside sandstone walls were covered with a mixture of cement and sand to fill all the holes and scars created over time. A second coat of stucco with color will be applied over the exterior (as soon as my shirttail relative raises a profitable crop on his land).

The small round openings, one over each door or window, are the only outside source of light in the attic or second floor. Inside, each opening is neatly boxed in and functions as an operable window.

Figure 2-2

This family home in France is over 200 years old.

The raised trim of stone around the door may or may not be redefined with a heavy finish coat of colored cement, to be blown on with a gun. The plaster gun is as recent an innovation in Europe as it is in the U.S.

Examples from commercial construction

As I have said, one need not go to Europe for examples of fireproof homebuilding. Commercial construction in the United States has always made use of fireproof techniques, such as framing with galvanized steel studs, heavy use of concrete, and so on. High-rise apartment and office buildings designed to bear heavy loads must, according to code, be built of more durable materials. More durable materials are usually fireproof. In many cases, all the architect need do is adapt these commercial methods to residential building to create a truly fireproof home.

All steel studs were used in the construction of this four-story apartment complex shown in Fig. 2-3. There was no danger of fire during the construction, and there will be no loss from fire of this noncombustible material after completion.

Figure 2-3

Four-story apartment complex with steel studs.

In Fig. 2-4, we see these fire-resistant materials at work in a small tract of houses in Southern California. Because land is expensive, lots were small, and units became two-story. All galvanized steel studs and trusses were used. This is an excellent example of how fireproof techniques can work in American-built homes and a great step forward in the use of noncombustible materials in the housing industry.

Figure 2-4

Steel framing in a small tract near Los Angeles.

Foundations

WHEN you build a house, you start with the foundation. The usual foundation, and the most economical, is the concrete footing, used for moderate-sized structures where ground conditions pose no unusual difficulties, such as ground fills or water conditions. Foundation for block, brick, or adobe will consist of two half-inch (1.27 cm) rebars continuous under all load-bearing walls. Of course, plans will vary according to terrain and frost conditions from north to south, especially if you wish to pour below the frost line.

For American homebuilders, a fireproof foundation comes as a matter of course. The concrete foundation, and to some extent even steel reinforcing, are traditional methods used in foundation construction in the United States. The Europeans, of course, also use concrete for the second floor, as well as for the roof. Our deviation from the European methods is one of the primary reason our houses burn.

In Fig. 3-1, you can see a section of a foundation with continuous lapped steel, typical of block below ground level. Number fours embedded in the foundation for every cell will continue through the last course of block, with horizontal steel every two feet (0.601 m) (Fig. 3-2). In some cases, you might need to make provision for gravel and seepage pipes along the block wall to drain off excess water. Any wall below natural grade would be poured solid with concrete mixed with pea gravel *(grout)*.

Columns

Piles are used primarily where surface soil conditions will not support a regular foundation. They are made of steel columns, or concrete columns driven into the soil or poured into caissons. This method might pertain to the residential construction when a house is cantilevered over a hillside for a view of the ocean or any scenic valley that may warrant the additional expense. The front of the proposed house will set on a concrete slab or piles that go into the soil to a depth sufficient to stabilize the building, and especially to keep it from sliding down, down, down.

Figure 3-1

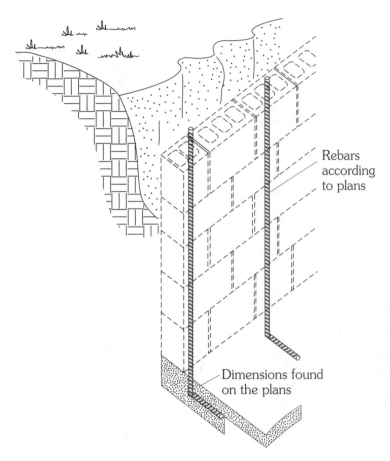

Rebars
according
to plans

Dimensions found
on the plans

Section of a foundation with continuous lapped steel.

Along the cliffs facing the ocean near Los Angeles, wooden columns
were at one time the choice to support a cantilevered west view.
Over the years, these posts began to disintegrate from dry rot
and/or termites. The supports were in serious danger of giving way
if too many people congregated in the post-supported room. In such
instances, regular inspections by the building department or an
engineer can identify the problem in time to have it rectified. Of
course, it's easier to avoid the problem altogether by avoiding wood
in these situations, and choosing more durable (and fireproof)
methods of support.

Figure 3-2

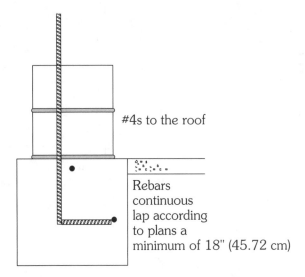

#4s to the roof

Rebars
continuous
lap according
to plans a
minimum of 18" (45.72 cm)

*Detail of a typical
section of foundation
with continuous steel
and uprights every two
feet.*

If columns are to support a building, you might find it necessary to install circular ties of reinforcing every so many feet (0.3048 cm) or inches (2.54 cm) surrounding the upright steel (Fig. 3-3). It's a good idea to look on your engineered plans for specifications. The columns should be poured with ready-mix or grout mixed on the job.

If people insist on living next to rivers, contractors need to start building these houses on columns such as the ones shown in Fig. 3-2. The columns then serve as stilts, placed approximately 15 feet (4.57 m) above ground and 15 feet (4.56 m) below ground (an engineer can calculate the exact proportion for the particular situation). If a house is built in this manner, it will merely shake when a chunk of ice from the spring thaw hits it—instead of collapsing into the river.

An architect or engineer can design various types of foundations for unusual conditions, all complying with building department regulations. Fewer homes would have been destroyed in the Landers and Big Bear earthquakes in 1992 had all the plans passed through the building department. I could tell from the piles of cement blocks shown on television footage after these quakes that steel reinforcing was not used. The lack of reinforcement is probably the reason the structure fell to the ground.

Figure 3-3

Typical column with reinforced steel.

According to engineered plans, columns should support any simple concrete structure you propose to build. After the foundation has been poured, set the assembled reinforcing over the rebars, and wire them together; then proceed building the columns (Fig. 3-4). When building in lowlands near rivers that might flood, you should make sure that houses are set on reinforced, poured concrete columns or walls. (This method would be much better than the bamboo stilts used in the tropics for such conditions!)

Figure 3-4

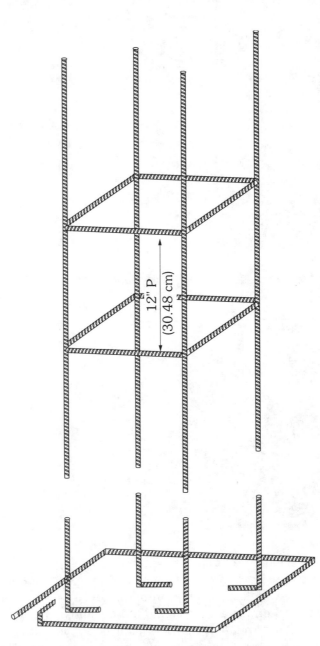

12" P
(30.48 cm)

Build columns for the foundation over reinforcing (in the foundation) and around steel reinforcing when in place.

Reinforced concrete slabs

Concrete is derived from portland cement, fine and course aggregate, and, in some instances, small amounts of entrained air with sufficient water for a pliable mix. Of all the building materials on the market, only concrete can be delivered to the job by trucks in pliable form where it can be poured into previously built forms for different designs. Concrete is used in high-rise apartments for separation of the different floors, where fireproof construction is required. You will find more detailed information about concrete in later chapters, as we discuss fireproof roofs. Here we are concerned only with the foundation for the structure and the slab for the ground floor.

Any time you have a row of columns, you need a webbing of steel in the foundation and upright steel running through the columns. Use concrete mesh reinforcing in the slab to keep cracks to a minimum and to hold the floor together. For brick or block structures, upright steel is placed in the foundation where it will continue through the roof.

Heating and air conditioning, in many cases, need to be placed underground before the slab is poured. If the job is big enough, you can pour the trenches, realign the upright reinforcing, and pour the slab the next day. But most jobs should be a continuous pour, especially where frost might separate the material at the joint.

A word about underground plumbing: Today, plastic pipes are often used instead of cast iron or tile. While plastic is a more combustible material than the older methods, plastic *underground* plumbing will never have a chance to burn. After all underground trenches, forms, rough electrical (in tubing, preferably), and plumbing have been inspected, concrete can be poured directly into the ground (Fig. 3-5).

For high-rise structures or even factories a soil test might be required. In this way you determine whether or not a sandy soil has sufficient compaction to support the proposed building, to prevent the structure from slumping in one corner due to poor soil conditions. The building department may give you a choice of

23

Figure 3-5

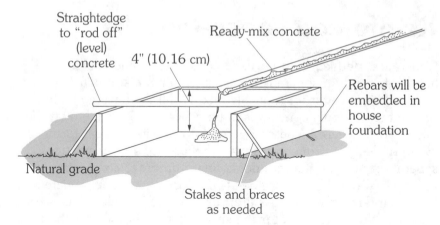

Straightedge to "rod off" (level) concrete

4" (10.16 cm)

Ready-mix concrete

Rebars will be embedded in house foundation

Natural grade

Stakes and braces as needed

Use ready-mix concrete whenever it is available.

deeper and wider footings or a soil test to determine the exact foundation requirement for that building.

Ready-mix concrete is really the way to go to get you in and out of this stage of construction. If you are building a public job, you might need to supply a hardness test. Take a plug from any wall and notify the ready-mix company of your requirements. In this way, you are assured a uniform quality that will meet the specifications stated on the plans and, ultimately, satisfy the inspector.

When pouring a typical slab, be sure to have at least one knowledgeable worker or subcontractor on the job, even if you are only pouring a garage floor. On any larger job, you should depend on an experienced crew. Most builders today rely upon troweling machines to take the place of several laborers. Unfortunately, run by an inexperienced worker, these machines can gouge great holes in the slab. If this happens, your slab will require a final hand troweling by a cement finisher if you want a truly smooth job.

Finally, before pouring, ensure that the basement floor is the same level as the proposed concrete slab for the house. Trench partially dug, with steel, will tie in with basement steel after earth is tamped around basement walls. Block layers who haven't had to contend with steel might find it difficult to lift and set in place each block above the steel sticking up in the foundation (Fig. 3-6).

Figure 3-6

A sight level will ensure the foundation is level with the floor.

Protecting the concrete finish

You should have a set of plans where the depth and width of trenches are defined. These will vary, depending on climatic conditions, and they are particularly important in frost-prone areas.

The surface of fresh concrete can freeze and will chip and flake off indefinitely thereafter. When there is a chance of frost, fresh cement, until it dries somewhat, will have to be covered with something. In dry areas, the most accessible material and the cheapest will probably be dirt, an inch (2.54 cm) or more.

Plastic is a good idea in rainy areas (Fig. 3-7) to cover the ground before pouring the slab to keep moisture from seeping upward. You might use visquene. However, keep in mind that you will have to wait for some time to trowel the slab if plastic prevents moisture from being absorbed by the soil. In these cases, you might want to add a setting agent to the mix.

Figure 3-7

Note the plastic covering in the right of the photo and the earthquake strap in the foreground.

Reinforcing is not always required, but it's a good way to ensure the slab doesn't crack. Note the earthquake strap caught under the lower rebar in Fig. 3-8.

An example here of a Midwest foundation is laid up of cement block with the plate bolted to anchor bolts embedded in the filled cells (Fig. 3-9).

During the heat of the summer, after the initial set of the concrete, spray the slab with a curing compound, or trowel it until it is hard enough to prevent small cracks from appearing. Surface checks in finish concrete have no adverse effect on the structural soundness of the building, but they become undesirable on a surface troweled by a journeyman cement finisher. In winter, moisture enters the small

Figure 3-8

Change visquene and reinforcing used under slab.

checks, expands and starts to chip away your finish surface. Small checks are caused by a rich mix of materials; a lean mix will not readily check, but it is difficult to reach a desirable, smooth finish.

After the first initial set, sprinkle the surface several times the first day and for several days thereafter. Again, you must cover the slab if rain falls when the surface has not set.

Concrete slabs won't burn. So far, the house is fireproof.

Figure 3-9

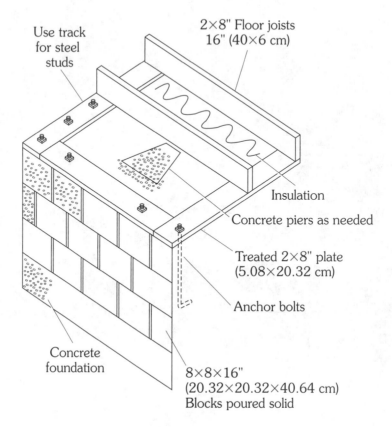

Use track
for steel
studs

2×8" Floor joists
16" (40×6 cm)

Insulation

Concrete piers as needed

Treated 2×8" plate
(5.08×20.32 cm)

Anchor bolts

Concrete
foundation

8×8×16"
(20.32×20.32×40.64 cm)
Blocks poured solid

Typical Midwest foundation plan.

More tips

All sidewalks around the house should have reinforcing bars from the slab extended into the sidewalk. Otherwise, after a few years, the house will separate from the concrete slab.

Concrete driveways should have felt expansion joists that will prevent the concrete from breaking and buckling in freezing climates.

Basements

A properly built basement is always fireproof. As discussed in the last chapter, you start with an adequate foundation, with reinforcing embedded at every eight or sixteen inches (20.3 or 40.6 cm). Many basements are built of cement block, setting the block over the vertical rebars in the foundation. Rebars should be embedded in bond-beam blocks every two feet (0.609 m) horizontally; all cells should then be poured solid.

In building basements, I usually see cement block being used (Fig. 4-1). This doesn't mean that the poured concrete walls are not used in many areas. In fact, I consider poured basement walls superior to cement block; there are no joints or small cracks for water seepage. It does however, in my opinion, take longer to build poured concrete walls because of the extensive work of assembling the forms.

Rebar safety

Rebars embedded in concrete and sticking up in the basement, such as the ones shown in Fig. 4-1, can be very dangerous. Workers have been known to fall on occasion, and a rebar can impale them. One man fell, and the bar went through his lungs and heart. The alert crew miraculously saved him by sawing off the rod at the jobsite and rushing the injured worker to the hospital, where doctors removed the rod and performed surgery.

Protective plastic coverings made of heavy rubber can be used to cover the bars and prevent such accidents (Fig. 4-2). These long, horn-shaped devices are placed over the upright bars while the crew is working above the dangerous area. The horn is blocked off at the top, so there is no danger of the rods puncturing anyone who falls into the hole striking a covered upright rebar.

You can also bend the bars slightly, if no more than half an inch (2.54 cm), so that they pose no danger to the workers above. When the blocks are laid up, bend them back into an upright position.

Figure 4-1

Typical block basement. Note unprotected rebar tops.

Figure 4-2

Safety covers for reinforcing bars help protect workers above from serious injury in the event of a fall.

Installing wiring

At the four-foot (1.22 m) level, after the horizontal steel is in place, install your half-inch (1.27 cm) tubing for electric along the center of the hollowed-out block. The tubing will have to be bent slightly to enter each outlet box.

Circle the room with tubing according to your needs and according to code requirements. Make sure of the minimum requirements, particularly if you are pouring a solid concrete wall. It will not be easy to add a box unless it is surface wiring. It is best to avoid tubing that will lessen the beauty of the room. An exception would be in factories where the wiring must be changed regularly according to current needs, as tenants move in and out.

You can pull wiring through the tubing embedded in solid walls at any time, even adding new wires as needed later. By all means, tape all splices in the tubing as well as the joints where the tubing enters the boxes. Otherwise, a small amount of juice from the grout might enter the tubing and plug the passage so you will have to abandon all your work.

Basement walls

Figure 4-3 shows deep leg track bolted to the interior basement walls. C-joists were used to span 20 feet (6.096 m) in the underground room. You can pull wiring through the openings in the punched joists. Use angle bearing clips as reinforcing, either screwed in place to the track and C-joists or welded in place. In crucial situations, an engineer might include specific requirements on the plans.

Figure 4-3

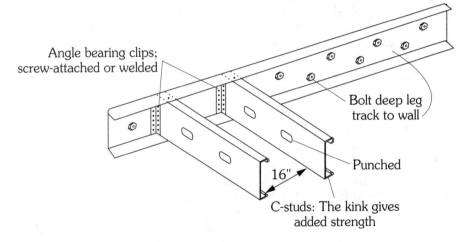

Angle bearing clips;
screw-attached or welded

Bolt deep leg
track to wall

Punched

16"

C-studs: The kink gives
added strength

Basement wall with deep leg track and C-joists.

You will have to paint exterior basement walls with a heavy coat or two of some type of plastic roof coating. (In Fig. 4-1, note the exterior walls painted with two coats of heavy roof coating applied with a paint roller.)

Where the climate is rainy, you can't waterproof the walls too much. Any painting of the interior of basement walls with waterproofing is an afterthought, and it will not keep water out nearly as well as plastering the exterior walls and troweling those smooth, topped by an asphalt coating. Cement is the only allowable material to be used underground when laying up block and plastering basement walls.

In the Midwest, many basements were at one time built of cement block, without reinforcing steel or solid grout. Eventually the walls pushed inward and the house was no longer setting on a foundation.

My uncle built a fashionable home when I was a young man. The block contractor had used no steel in the basement walls, nor had he filled any of the cells with grout. Later, the outer basement walls started to cave inward. After my uncle and aunt died, the land was sold. With a few termites in the walls of the house and the basement walls caving inward, the people who bought the house ended up tearing it down and starting over. What a waste! Just as bad as having your house burning down. The more rigid building codes of today require upright steel with at least one rod in every block and horizontally placed steel at intervals of every two feet (0.609 m), then poured solid. As a result, homes today are unlikely to suffer a fate similar to my uncle's house's fate.

Even the area between the 2×10 ceiling joists (5.08×5.08 cm) in the basement should be insulated, regardless of whether rigid building codes are in force for that phase of the work. With a screw gun, I fastened rib-lath to the basement ceiling in my home and applied two layers of cement plaster. If you do this, attach the metal through the large ribs so the screws never become detached in an unforeseen disaster. If there is only regular webbing for attachment, use a washer over every screw for a wider area of security.

In a quake, metal lath becomes detached from the ceiling joists and will hang down. The great weight of the material is left in one solid piece, which can pull the entire ceiling down. Washers for a wider area of attachment prevents the ceiling from becoming detached. Such washers used to be called *schoolhouse* washers at a time when many schools were built in California. Wanting the best from the construction industry for the children, the builders used metal lath for a plaster base over wood studs and ceiling joists. Today, they use metal studs and ceiling joists with metal screws to fasten the metal lath.

After the two layers of cement plaster, I added white cement to acoustic and shot the basement ceiling. On the upper floor, I covered the steel joists with corrugated metal and screwed it to the joists in a few places so one could safely walk on the metal (Fig. 4-4). I then

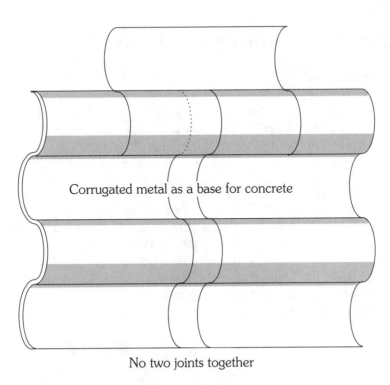

Figure 4-4

Corrugated metal as a base for concrete

No two joints together

Corrugated metal flooring.

had a welder tack the metal to the joists according the schedule on the print.

When dealing with corrugated metal flooring, buy as long a length as you are capable of handling. Also, have everything ready for the welder so the welder hasn't nothing to contend with but welding. Break any joints when applying metal to the joists, just as you would when lathing.

All joists should be of a metal soft enough so self-tapping screws will penetrate; otherwise, you will have to predrill all holes before attaching metal with screws. Specifications for joists should be on the plans. C-studs 10-inch-x-20-feet (2.54×5.08 cm) work fine. The extra kink in the stud to form the C gives added strength to the joist.

Basement stairs and floors

Over the metal on the basement floor in my house, I attached one-inch (2.54 cm) 20-gauge chicken wire and applied two inches (5.08 cm) of grout, (concrete mixed with pea gravel). Without the chicken wire, the concrete cracks too readily, especially with only two inches (5.08 cm). It really doesn't matter too much if carpeting is added, but any other floor covering should be applied over a floor that doesn't crack.

For the stairs, use C-studs or I-beams, according to the design. Over these, apply metal pans. The welder will set the beams in place, then weld the triangular metal to that and the pans to the triangles, which levels the treads. Pins can also be shot into the wall to which a railing is attached. If the plans are not clear (or not available), allow a good welder to build the stairs (Fig. 4-5). Use no wood—especially if you are building your stairs outside close to a river that might rise some spring night when the snow melts.

Figure 4-5

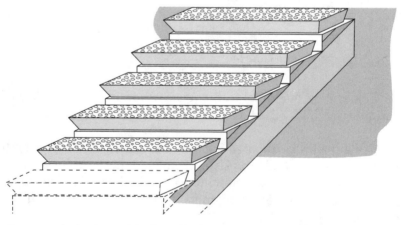

Stairway bolted to a block wall, the bottom of the I-beams embedded in concrete. Triangular steel supports the treads, which are pans filled with concrete.

Inside the house, you might want to use the same procedure, and even pour the concrete into the pans yourself instead of buying those already filled. If you want to apply a rug over the hard concrete and steel, you might need a 2×2 wedged in back of each tread so the rug can be fastened with tackless strips.

36

5

Fireplaces

YOU have only to visit the foothills of California after a runaway fire to come to the conclusion that this country needs to build with brick or block. You might get out of your car and survey the damage. Fireplaces are still standing, alone, the only remains in the wake of the fire, as if guarding the blackened hillsides once green and graced with beautiful homes.

In this country, fireplaces are always left standing after a fire because they are the only things built with fireproof materials in mind. The need for a noncombustible fireplace is obvious to everyone: You could scarcely light a fire in something made out of wood or other combustible material and expect it to survive. As a result, fireplaces are made of brick and block, and lined with ceramic or cement liners, and all sorts of codes are in place to protect the rest of the house from the fireplace danger. Of course, little thought is given to the inherent combustibility of traditional homebuilding materials.

The pile of rubble in Fig. 5-1 would probably be standing today had the house been built as well as the fireplace. Even after the recent fires in Oakland, they are rebuilding with wood, creating $200,000 and $300,000 homes that consist of a pile of kindling nailed together with an application of stucco and wallboard.

Extra protection

A metal fireplace liner is not nearly as successful as a fireplace built by a journeyman mason and lined with firebrick. I once had a very successful fireplace built with no liner at a reasonable price. Then later I had a fancy fireplace built around an expensive liner. This second fireplace allowed smoke to accumulate in the house, while the first fireplace gave no trouble with smoke.

A spark arrester or screen over the flue is important as a fire protection for the surroundings everywhere, especially where vegetation is plentiful and dry (Fig. 5-2). On my most recent fireplace, the inspector wanted the screening flush, but wind whistled down the chimney and filled the house with a thin film of smoke.

Figure 5-1

Only the fireplace remained after a fire gutted the rest of this house.

My solution was to build a gable shed over the opening, blocking out the wind (Fig. 5-3). I had to figure out a means of anchoring the material so it didn't slide onto anyone. To do this, I had half-inch (1.27 cm) holes drilled into the brick and six-inch (1.52 cm) rebars driven into place. If the flat slab is cast and placed over the fireplace, use chicken wire in the casting. This solution worked fine, and the house no longer smells smoky even in the worst wind.

Figure 5-2

Double fireplace with spark arrestor.

Proper attachment and reinforcement

I learned firsthand from the Silmar earthquake that a fireplace had to be properly attached to a house. I drove around taking pictures of quake damage. Those fireplaces where code regulations were followed to the letter were scarcely damaged. In the case of the house

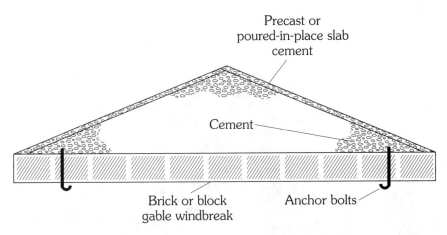

Figure 5-3

This gable obstruction against the wind was actually built as an after-thought for the fireplace above.

in Fig. 5-4, reinforcing rods were placed in the foundation or pad especially poured for the fireplace. The rods were continued to the top of the fireplace, but the attachment to the house was not adequate or missing, and this allowed the fireplace to lean out. Probably the brick contractor failed to provide the straps.

You can buy various sets of plans on how fireplaces should be built and reinforced. In Fig. 5-5, the liners to form the flue come in either clay or cast cement and are about 18 inches (0.4572 m) high, with about a 14-inch (0.3556 m) oval (or larger). These are laid up with the fireplace wall. Then, grout is poured progressively around the liners and between the outer wall of the fireplace that contains the reinforcing steel. The rebars of the metal straps are embedded in this pour.

Figure 5-5 shows a typical strap used to tie the fireplace to ceiling joists. The end of the strap with reinforcing bars is embedded in the fireplace. Grout is then poured around the reinforcing, extending from the foundation through the finished top of the fireplace. It's best to catch the rebar part of the anchor strap behind an upright rebar of the fireplace. The carpenters can then later bolt the straps to the ceiling joists. Bolt holes are provided in the straps.

Figure 5-4

This fireplace failed to withstand the shock of an earthquake.

Figure 5-5

1¹/₂" (6.35 cm) Metal strap with ¹/₂" (1.27 cm)
anchor rods to fasten fireplace to the house.

You won't have a house fire caused from a fireplace built to this cross-section. Note the horizontal and upright reinforcing bars.

Figure 5-6

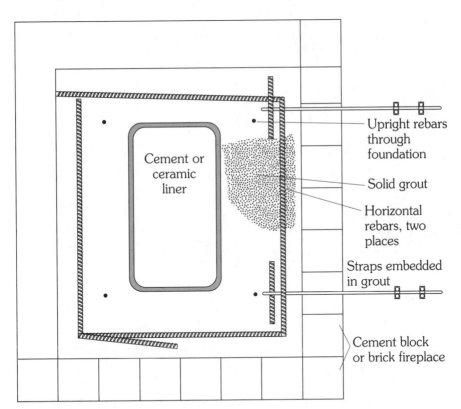

Cement or
ceramic
liner

Upright rebars
through
foundation

Solid grout

Horizontal
rebars, two
places

Straps embedded
in grout

Cement block
or brick fireplace

Typical strap to tie the fireplace to the ceiling joists.

Use two straps as a minimum, more if the plans indicate. You can buy all the required items from your local material yard that supplies brick products.

Many tract homes today have a fireplace cast and hauled to the jobsite and set in place. These fireplaces, too, have to be properly fastened to the house. Bolts and or straps are cast with each unit and coordinated with attachments to the house. This type of fireplace is never elaborate, but taken from various molds at the factory. They do serve their purpose, if the hearth will accommodate a log of sufficient size.

6

California-style block homes

DUE to the frequent shaking of the earth's surface, California has instituted building requirements that far exceed the demands in other parts of the nation. The use of steel in block walls renders them rigid when those cells are poured full of concrete. The walls will then bend slightly; the building will shake but remain standing through an earthquake.

These rigid building codes need to be extended to tornado- and hurricane-prone states where they are not already in effect. A wood-frame house in Florida will readily blow away, while a wood-frame house in California will withstand a quake better than any other material. During a quake, the lumber might squeak, but it will allow some leeway while the house remains intact, with only plaster damage. Of course, after a shake, the wood-frame house could very well burn from an electrical fire. I have ventured onto the fireproof construction program and have turned to the block-wall structure rather than wood for this reason.

Building with block or brick has been standard procedure in many Eastern, Southern, and Northern states. The earthquake districts have been the exception; the reinforcing perfected to hold the structure together during the process of a good shake meant additional cost, so the price of the house becomes higher, which discouraged many people from building with block as they do in other parts of the nation. Now it is time to consider fireproof material as insurance against rebuilding.

The do-it-yourself homebuilder is more capable of working with lumber than with block. But, when building on your own, it is sometimes feasible to hire a subcontractor or journeyman to do any phase of the job that proves difficult.

The basics

For a fireproof structure, cement blocks are probably the cheapest when compared to brick or adobe. Because they are porous, however, they leak with any appreciable amount of rain so they might have to be plastered outside or waterproofed in some way; in some cases, both applications might be necessary. Three- or four-foot (0.91 or 1.22 m) eaves or an overhang will help to keep water off the walls.

The foundation detail shown in Fig. 6-1 is designed to withstand a shock from an earthquake and would be ideal in hurricane districts. The foundation would require two pieces of continuous steel in the trenches, one at the top and one at the bottom. The upright bars would continue to the roof and through the parapet wall, if one is required. Every third cell containing a rebar would be poured solid. The slab sets six inches (15.24 cm) above natural grade.

Figure 6-1

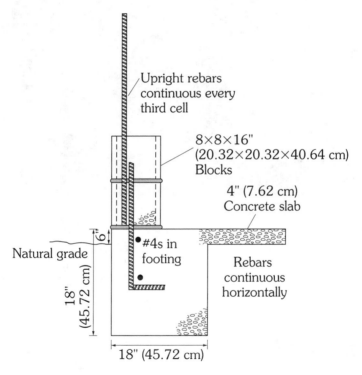

Typical foundation for a block building.

The California-style building program using reinforced block is shown in Fig. 6-2. States lagging behind in the use of reinforcing steel in homebuilding need to upgrade their building codes. To attain the beautiful surface shown here, the surface is crushed off by a machine after the blocks have hardened for a determined amount of

Figure 6-2

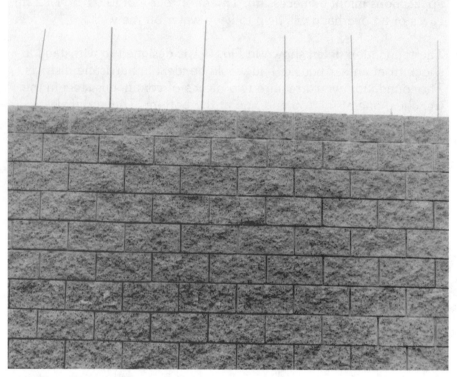

Wall of California-style block.

time. All rebars out of place will be realigned before grout is poured into the cells containing reinforcing. A neat block job takes no longer for a journeyman mason than a sloppy job performed by a less-experienced worker.

Walls and ceilings

The typical block wall is eight inches (20.32 cm) thick, generally with two cells or air spaces to the block that act as insulation. Blocks come in various sizes, 8×8×16 (20.32×20.32×40.64 cm) being the most commonly available, but the thickness or the width will vary. The 8×8×16s (20.32×20.32×40.64 cm) are used in mainly commercial buildings. The shallower depths, 4×8×16 (10.16×20.32×40.64), have more grace or beauty and are used generally in residential construction. Unless the blocks are going to be plastered, I recommend using the larger blocks, which tend to be cheaper.

In most areas in the United States, blocks are laid up without contending with steel (Fig. 6-3). Bolts or angle iron are embedded in the top of the wall to fasten down the roof structure. Depending on the plans, insert the anchor bolts into the bond beam before pouring the grout, either on top the wall or alongside it. Place the bend in the top bolts under a horizontal reinforcing rod or in back of an upright rod for the side wall. The roof structure can then be bolted to the anchor bolts within the wall. The ceiling joists would be fastened to the bolts within the walls. Before the grout hardens, wiggle the bolts so they are true and straight, ready for any attachment (which will be, hopefully, galvanized steel members, rather than wood framing).

Figure 6-3

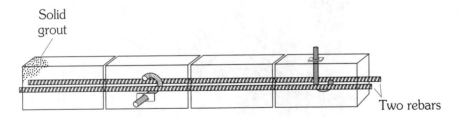

Solid grout

Two rebars

Two ways of bolting framework to the building.

The plans should be engineered and the house built to contend with the various destructive forces of nature. Figure 6-4 actually shows a sloppy block job, but you can see that every third cell will contain a half-inch (1.27 cm) reinforcing bar. These cells will be poured solid with grout (check plans for exact proportions). Here, a steel door frame is fastened to the wall by brackets that come with the package. The brackets attached to the steel door frame become embedded into the grout during the pour. In the west, cement block structures, because of rigid steel requirements, are found mostly in commercial work. Of course, the fireproof nature of the product makes it an ideal one for fireproof residential construction.

Figure 6-4

Although this is a sloppy block job, you can see that every third cell contains a rebar.

Steel rebars

Because California building codes are more strict for block, brick, or adobe, steel must be added to render these structure more rigid. With any shifting of the earth's crust, the steel will bend slightly without allowing the building to topple.

Figure 6-5 shows two types of bond beam blocks and how steel is used horizontally. Steel in the bond beam blocks encircles the building every six courses, with laps of at least 15 inches (38.1 cm) wired together. This ties the building together as a unit so it won't fall apart during a good shake or wind. This row of blocks is then poured solid with grout. The cells without upright blocks are blocked off so the whole wall is not poured solid.

Figure 6-5

Two types of bond beam blocks, with proper use of steel indicated.

There are definite codes pertaining to steel when combined with cement block walls. Rebars (reinforcing bars) are graded in increments of one-eighth inch (6.4 mm). A #4 is a half-inch (12.7 mm) bar, a #5 is a five-eighth-inch (32 mm) bar, a #6 is a three-quarter-inch (19.2 mm) bar, and a #8 is a one-inch (2.54 cm) bar. The rebars start in the foundation where they are placed prior to pouring the concrete. Every third cell of two cell blocks will contain a #4 reinforcing bar, or a #5 every 32 inches (0.8128 m). The upright bar is bent into an L-shape of about six inches (15.24 cm) to keep it from pulling out of the concrete during an earthquake or tornado, like a toothpick pulled from a cake (Fig. 6-6). Again, these specifications vary, according to plans as designated and approved by the building department.

Figure 6-6

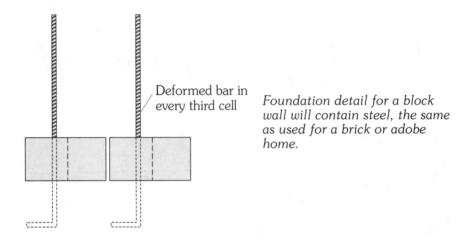

Deformed bar in every third cell

Foundation detail for a block wall will contain steel, the same as used for a brick or adobe home.

The cells with the bars are then poured solid with grout, while the other cells are blocked off with a slice of metal lath (Fig. 6-7). Another method of blocking off the cells is to place the blocks on a layer of tar paper on the cement slab, dropping a trowel full of mortar in each cell to be blocked off. The next day, after the trowelful of cement hardens, incorporate the blocks in the wall as needed.

General block construction

You can find the designated proportions for mixing cement on the plans, but a good mix is comprised one part cement to two-and-a-half parts pea gravel and two parts sand. Obviously, if a ready mix truck brings the material, you have little worry about the mix, provided you have informed the company of the required mix when you order grout or have shown them the specifications sheet.

Figure 6-7

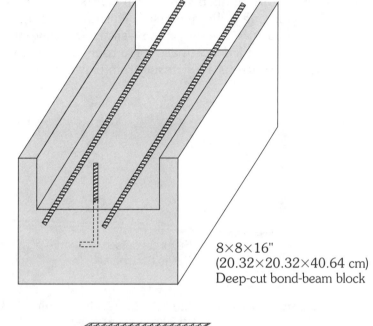

8×8×16"
(20.32×20.32×40.64 cm)
Deep-cut bond-beam block

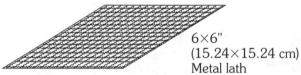

6×6"
(15.24×15.24 cm)
Metal lath

Block walls using 8-×-8-×-16 block will be grouted or blocked off with metal lath.

Every four feet (1.22 m) in the course of a typical eight-foot (2.44 m) wall of any structure, lay a course of bond beam blocks around the entire building, blocking off cells without the rods as previously mentioned. Then place two #4 rebars where provision has been made in the bond beam (refer back to Fig. 6-5).

Next, lay the blocks accordingly up to the eight-foot (2.44 m) level, 12 or 13 rows of 8-×-8-×-16 (20.32×20.32×40.64 cm) blocks in all. (You might want 13 rows to avoid clipping your ceiling slightly from the standard eight-foot (2.44 m) height.)

The last two courses are of deep bond beam blocks to accommodate two pieces of steel in one or both courses. This forms a bond beam around the entire structure and keeps the walls from falling out during an earthquake, which would allow the roof structure to collapse inward. It also keeps a hurricane from blowing your building down (Fig. 6-8).

Figure 6-8

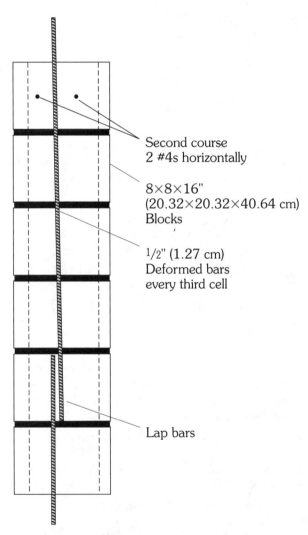

Second course
2 #4s horizontally

8×8×16"
(20.32×20.32×40.64 cm)
Blocks

1/2" (1.27 cm)
Deformed bars
every third cell

Lap bars

Row twelve gets two courses of #4 steel and will be poured solid (as is row six).

You will need to use lintel blocks over all openings to accommodate reinforcing—two #4s in each of two courses above all openings (Fig. 6-9). These will be poured solid to support the superstructure of block. Lintel blocks should also have a groove or slot to accommodate certain types of windows. As the windows are later lifted into place, a sash block with a water drip will be inserted. And, of course, caulking will be necessary around all openings.

In Fig. 6-10, you can see that provisions for plumbing have been made (in this case, for a wall-hung toilet that will facilitate scrubbing of the bathroom floor). Figure 6-11 shows an attached garage that was built first. Notice the steel in the trench and at the two-foot (0.609 m) level to tie in with the walls of the house.

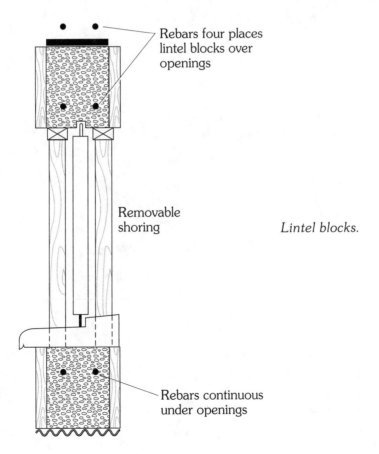

Rebars four places lintel blocks over openings

Removable shoring

Rebars continuous under openings

Lintel blocks.

Figure 6-9

Figure 6-10

Provisions for plumbing are being made.

Figure 6-11

Attached garage with steel in the trench and at the two-foot level to tie in with the walls of the house.

Ceilings and roofs

I built my own fireproof home with 8-×-8-×-16 (20.32×20.32×40.64 cm) blocks, using one 6×8×8 (15.24×20.32×20.32 cm) and one 8×8×8 (20.32×20.32×20.32 cm) whenever I need to leave a slot to accommodate a C-stud or joist (Fig. 6-12A). Punched holes in the joists were aligned so rebars could be threaded through the opening. This became a bond beam, which was then filled with grout.

Allow another two-inch (5.08 cm) vacancy if you plan to place rafters side-by-side with ceiling joists. Weld the joist and rafters together as insurance against high winds. This row of blocks with the embedded steel and rebars is poured solid (Fig. 6-12B) and becomes as cyclone-proof a roof assembly as is possible. These days, engineered trusses might be required, perhaps eliminating this method of joist attachment to block. These details have to be engineered for each individual plan until the method is recognized as standard procedure.

Figure 6-12

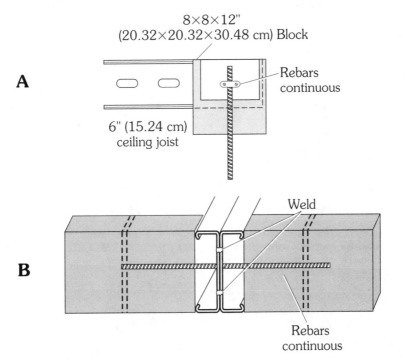

(A) Rebar through foundation every two or four feet. (B) Ceiling joist and rafter welded back-to-back. Pull rebars through rafters and punched joists.

For nearly flat roofs, punch holes in each ceiling joist to line up and thread rebars through (Fig. 6-13A). Plaster the overhang with Z flashing in place. Punch holes through the rafters, too, and thread the rebars through the holes in both the joists and the rafters. If the holes in the material do not match, burn them through with a welding torch (Fig. 6-13B).

In Fig. 6-14, you can see ceiling joists at 16 inches (40.64 cm) off center, embedded between cement blocks. These are two pieces of half-inch (12.7 mm) steel pulled through the punched holes and bond beam block poured solid with block. Note the screws for fastening the metal lath ceiling have neatly penetrated the C-joists. Nearly all steel trusses in homebuilding are now placed two feet (0.609 m) OC, with the wall studs placed accordingly—a savings in material that might sway some homebuilders to using steel instead of lumber.

Iron brackets are sometimes used to bolt rafters to a block wall (Fig. 6-15). If brackets aren't readily available, you can have an ironworker or blacksmith fashion them for you. Or your architect or engineer might have a pet idea for an equally effective attachment.

Figure 6-13

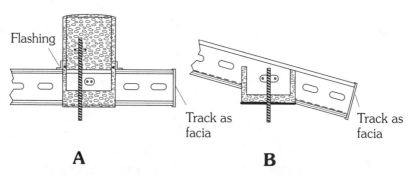

Flashing

Track as facia

Track as facia

A   **B**

(A) Nearly flat roof with punched joists with overhang. Pull rebars through punched holes, providing opening with welding torch when necessary. (B) Rafter and ceiling joists back to back.

Figure 6-14

Ceiling joists at 16 inches OC, embedded between cement blocks.

Figure 6-15

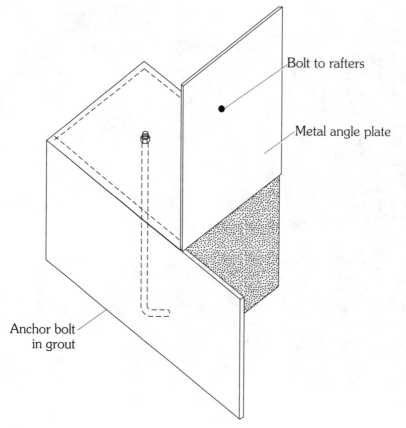

Bolt to rafters

Metal angle plate

Anchor bolt
in grout

Be sure to provide some way to fasten the rafters to the block wall.

Plans

After you study steel requirements for a block building in the earthquake-prone or windy states, you can readily see why so few block houses are built. Yet constantly building with wood adds to vulnerability—in creating fire hazard, as well.

Only a licensed engineer can put his stamp of approval on a building requiring brick, block, poured concrete, or adobe block to obtain a permit. The stamp doesn't mean you will get an immediate approval from the building department. They, too, have their plan checkers and will charge a fee to check the engineer's calculations to discover what they think are changes that might need to be incorporated into the plans. The engineer has to prove by his calculations that whatever one proposes to build will withstand an earthquake of a determined magnitude, or perhaps a 150-mile-per-hour (241.35 kilometer-per-hour) wind.

In some states, steel is not required in either brick or block buildings. But, I maintain, anyone planning to build in a hurricane- or tornado-prone district should promise himself that he will follow the same regulations in building with block, brick, or adobe as one does when dealing with earthquakes. There would be fewer buildings flattened in a twister—and, ultimately, fewer destroyed by fire.

Brick

FOREST and brush fires seem to be nature's way of renewing itself by cleaning out all the dead wood and starting over again. But reasonably new $200,000 and $300,000 homes are hardly dead wood, so they shouldn't be wiped out—unless you cooperate by building a wood-frame house. Such was the case with the Oakland fire in California in 1991. The brick fireplaces were all that remained standing, which really speak for themselves as to the materials recommended for a fireproof house.

Brick homes are built similarly to block houses, with steel reinforcing in the walls. Bricks are generally four inches wide and laid up in two courses with a four-inch hollow space in the middle for reinforcing and grout (California requirements). The dead air space in the middle becomes insulation in most areas, but grout poured around steel is essential in earthquake-, hurricane-, and tornado-prone districts.

Brick building is nothing new in the Northeast and, to some extent, in the Midwest, but it has not been used on the West Coast because of earthquakes. Gradually, more brick houses are being built in California as homebuilders become familiar with the incorporation of steel to reinforce brick homes.

Reinforcement

The steel starts in the foundation with a rod every two or four feet perpendicular, and then a rod every two or four feet horizontally. Figure 7-1 illustrates the recommended cross section of a foundation incorporating two pieces of half-inch (12.7 mm) reinforcing bars, one at the bottom and one at the top of the trench. (Of course, all dimensions are based on your engineered plans.) Keep in mind that a dwelling of more than one story requires heavier steel, which should be readily available. Under normal conditions, the house would be six inches (15.24 cm) above natural grade, as shown. This grid of reinforcing is tied together and inspected so grout can be poured continuously as work progresses.

The bond beam is fabricated similarly. Embed bolts as needed into the poured concrete for fastening down the roof assembly. Or the

Figure 7-1

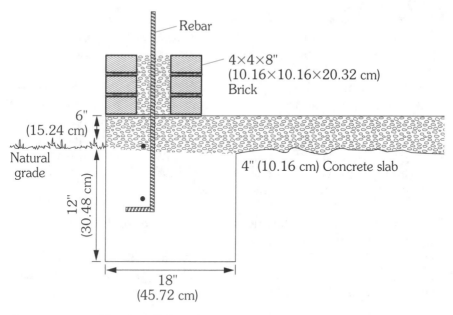

Cross section of brick wall foundation.

bolts might be allowed to protrude from the inner side of the walls for tying in the roof and walls. More about tying in the roof is found in the previous chapter's discussion of block walls. (Fig. 6-1.)

Brick is not the best insulation against cold or heat, but they can be furred out and insulated and they are a sturdy support for a fireproof roof, which in many cases is several times heavier than the frame roof.

Brick as veneer

Brick is sometimes used as a veneer over wood framing. This method has been one of the most successful in the Midwest. Bricks by themselves do not supply a warm wall, but they provide a substantial outer facing for a house and will last a hundred years. Using only one layer of brick instead of two makes a substantial home—one that would be fireproof except for the wood studs. Of course, the shift to steel is an easy one.

You start by building a frame house as you normally would, then lay the brick veneer next to a layer of tar paper attached to the 2-x-4 (5.08×10.16 cm) framing, using 15-pound (6.81 kg) felt or heavier. Leave an airspace between the brick and paper as a form of insulation. Attach the asphalt paper to the framing as you lay up the brick to keep the wind from blowing it away. Don't depend on the brick keeping out precipitation; your building must be watertight. The asphalt product must lap over the foundation or slab to drain away any and all rain. The paper has to be nailed to wood framing or screwed in place with steel studs on steel framing, one layer overlapping the other, so it sheds water like shingles.

Because the use of steel furring is coming to the front, 2-x-4 (5.08×10.16 cm) or 2-x-6 (5.08×15.24 cm) framing can start with exterior walls of steel studs—20-gauge should suffice for a bearing, fireproof wall for brick veneer. Six-inch (15.2 cm.) studs will, in many cases, support a tile roof.

On class A-construction, the workers will tell you that steel furring will gradually replace lumber, leaving the forest products to the owls. The vast number of school buildings and high-rises are now fireproof—constructed of steel furring in many instances, with some type of exterior fireproof veneer wall covering fast becoming the norm. This method of construction has recently crept into residential construction, leaving those buildings with fireproof walls.

As a homebuilder, don't settle for cost quotes that are twice as much if steel furring is used instead of wood. Get another figure from your subcontractors. Some framing contractors are anxious to learn the new trend. Some will take the job for wages just to learn. (On June 1, 1994, the material dealers in my area told me that steel studs were now cheaper than wood.)

Brick should be tied to steel to keep the brick from toppling in earthquake- and cyclone-prone districts. You can do this by fastening waterproof paper and stucco netting over the steel, shingle-style, using a screw gun. Then cover the building with a half-inch (12.7 mm) of cement plaster, scored. When you lay the single row of brick up against this surface, use plenty of mortar so the bricks adhere to the stucco wall.

Figure 7-2 shows brick ties used to fasten veneer to steel or wood studs. Fasten the tie to the steel studs with screws, then bend them into the next row of brick, applying mortar to keep the veneer from pulling loose in a wind or quake.

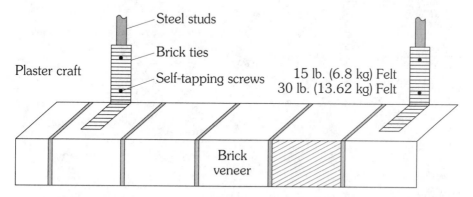

Figure 7-2

Brick ties used to fasten veneer to steel or wooden studs.

Paper-backed wire is more expensive than plain wire, but it's easier to apply over steel studs. Lap the upper sheet of material over the excess of paper of the lower sheet to keep out the rain, like shingles. Provision is made in the manufacture of the wire, so the netting is lapped a minimum of approximately one-and-a-half webs to keep the stucco from cracking (Fig. 7-3).

The lather or bricklayer should apply the paper-backed wire progressively as the bricks are laid up, to keep from scaffolding twice. I recommend using wafer-head Streaker screws or Grabber Phillips-recess screws attached to 20-gauge or 18-gauge steel studs. (Table A-8 shows screw sizes.)

In a steel-frame house, you can use 2-x-4 or 2-x-6 20-gauge galvanized steel studs (or the gauge designated on your particular set of plans).

In the west where quakes are a regular occurrence, the attachment of K-lath or some type of netting might be required over, we hope, steel studs. Metal lath attachment could be with wafer-head Streaker screws, 8-gauge by ¾-inch (19.5 mm), #2 Phillips recess to 20-gauge steel studs. The Grabber line has a similar screw: 8 gauge, ¾ inch (19.5

Figure 7-3

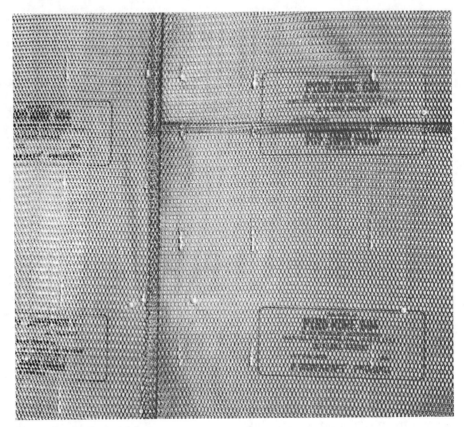

Paper-backed metal lath.

mm) attaches metal lath or paper-backed material to steel stud walls. These are applied with an electric drill with Phillips attachment.

The exterior wall covering is then plastered the first coat and scored before the brick are laid up, or the lath is plastered as the bricks are laid up. This attaches the brick firmly to the wall so they won't later lean away (or, in the case of a tornado or earthquake, they won't pull away).

In the west, plastering may not be deemed sufficient attachment of veneer to steel studs. Corrugated metal ties are secured in place with the same screws so the walls don't fall out from a movement of the earth or a tornado.

K-lath has metal stiffeners embedded within each sheet to facilitate plastering. With the application of regular stucco netting, the wire for stiffeners is attached to the wood framing. You can't do this with steel, so use paper-backed K-lath or equal, or paper-backed metal lath.

Figure 7-4 is not brick veneer, but a separate wall. In most localities, the inner brick wall might stand separate from the outer brick wall, but not where quakes or high winds are prevalent. In many locales, even the furring could stand separate from the brick walls. As mutual

Figure 7-4

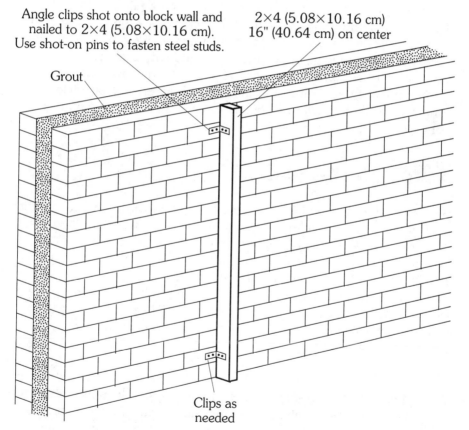

Angle clips shot onto block wall and nailed to 2×4 (5.08×10.16 cm). Use shot-on pins to fasten steel studs.

2×4 (5.08×10.16 cm) 16" (40.64 cm) on center

Grout

Clips as needed

Outer wall of brick, separated by grout from the inner wall.

support against a major shake or wind, the furring should be fastened to the outer wall with angle clips, perhaps shot on with pins or some equal method of attachment. Embed bolts or rebars in the wall to divide the rooms.

Energy efficiency

According to the Brick Institute of America, bricks's energy efficiency is universally recognized, proven over years of actual performance as a building product, and demonstrated time and again in research studies. It is common knowledge that brick helps keep the interior of a house warm in winter and cool in summer, while saving on energy costs. In general, we've understood how and why brick is such an efficient thermal comforter. Research projects, however, have probed more and more deeply into the thermal characteristics of brick, and findings provide important evidence of brick's energy advantages from the standpoint of mass or weight.

Through the walls

Builders and architects have become concerned with how much heat flows in and out of buildings through the exterior walls and how heat gains or losses are affected by the type of material chosen. Traditionally, such thermal effects have been measured in terms of U-values (the level of conductance of heat through a wall), and of the temperature differential from one side of the wall to the other. The calculations have proven that there is a steady heat flow of heat through the walls, a process occurring in a fairly constant manner throughout the day.

But there are daily and seasonal temperature fluctuations outside, so that the temperature gradient across the wall system is not constant. Other key variables are also important: the specific heat and surface absorbency of wall materials; surface color; the type, position, and thickness of the thermal insulation; the solar orientation of the building and the degree of solar radiation; and wind velocity. All these factors affect the true performance of the wall and its material composition.

Traditional analysis has before ignored the effect of mass, heaviness, thickness and how mass has influenced the heat flow through the walls. Considering these factors, heat transmission through building walls is clearly not static, but a dynamic process that must be factored into heating and cooling calculations. When mass was considered, brick energy advantage was seen to be considerable, perhaps a 30 percent savings over similarly insulated wood construction.

Passive solar heat

The mass of a brick wall makes a significant contribution to the energy efficiency of basic heating and cooling systems. But brick mass takes energy conservation many steps further when it is also applied to a passive solar heating system. It slows the absorption of heat and later slows the release of heat.

In a passive solar system, an interior brick wall collects and stores solar energy during the day, and then releases and distributes the heat during the evening hours when it is needed. Mass is a key element in solar storage; the density or heaviness of the brick retains heat and slowly and evenly releases it.

True passive solar heat absorption is a nonmechanical technique; there are no moving parts. The house should have a southern orientation, large windows facing south and minimal glazing on the north. With brick walls and floors for solar storage, the energy savings are substantial, even in wintry and cloudy locations.

Since brick keep houses warmer in winter, it reverses the process in summer and keeps the same houses cooler. That is the nature of mass weight in pounds of not too dense a material such as brick.

You can add to the energy efficiency of brick, the beauty, durability, freedom from maintenance, fire resistance, and resale value, and the net result is desirability and salability of the brick home, something to warm a homebuilder's heart in winter and keep his temperament cool in summer.

There is a never-ending amount of clay all over the world for the manufacture of brick. This should make the product cheap and everyone should take advantage of the use of brick in homebuilding.

Brick will give you an R-value of 0.20 per inch (2.54 cm), 0.80 for a four-inch brick-veneer wall, or 1.60 for an eight-inch (20.32 cm) wall. The R-value of grouted clay block is 1.60. Sand-plaster interior walls are 0.11. The R-value of many existing homes is 2.04—very bad. Nearly all should have a retrofit insulation job. This R-value might suffice for warmer climates, but it would definitely have to be furred out in colder climates.

Brick exterior can be combined with block interior. In this case, use one-inch (2.54 cm) stripping inside, and install wallboard for a wall covering. If you don't plan to grout the brick, use vermiculite. Figure 7-5 shows a cross section of a brick wall with vermiculite poured between the walls. Metal ties fasten the outer and inner walls together. A grid of ½-inch rebars (12.7 mm) allows work to be performed without waiting for further inspections.

Four inches (10.16 cm) of grout adds to the R-value of brick, increasing it to 0.44. Cored brick increases the R-value to 0.80.

Fill the four-inch (10.16 cm) space between the two rows of brick with vermiculite and increase the R-value to 9.51. This is feasible only with new construction, where steel and grout is not required such as in earthquake-prone districts.

Figure 7-6 illustrates how you can apply one-inch (2.54 cm) hat furring, shot to a brick wall to provide space for insulation. Attach the wallboard to the hat furring with self-tapping screws.

A typical section of a brick wall with reinforcing and grout is shown in Fig. 7-7. Anchor-bolt in place for the top plate or track, making sure that the anchor section of the bolt is caught under a rebar. Wallboard over the furring, increasing the R-value to 0.45.

Figure 7-5

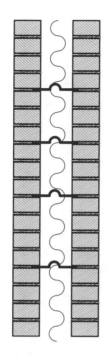

*Metal ties fasten the inner
and outer walls together, with
vermiculite in between.*

Figure 7-6

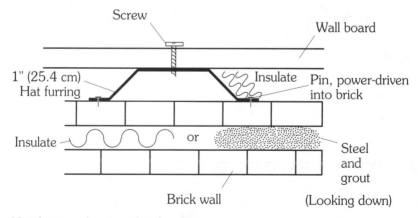

Screw

Wall board

1" (25.4 cm)
Hat furring

Insulate

Pin, power-driven
into brick

Insulate   or

Steel
and
grout

Brick wall   (Looking down)

Hat furring shot to a brick wall.

Figure 7-7

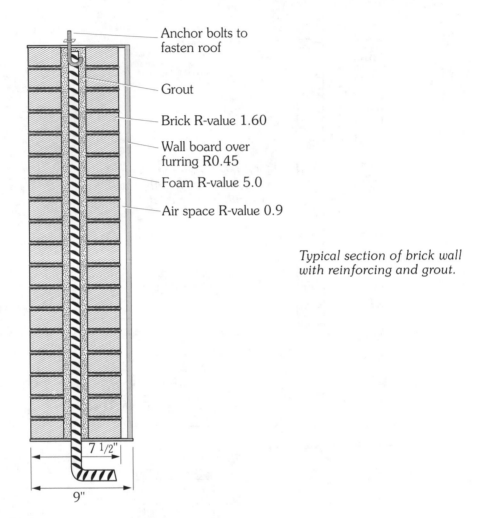

Anchor bolts to
fasten roof

Grout

Brick R-value 1.60

Wall board over
furring R0.45

Foam R-value 5.0

Air space R-value 0.9

*Typical section of brick wall
with reinforcing and grout.*

7 1/2"

9"

Interior partitions

At one time, clay tiles were used more extensively in the United
States for partitions than they are today. This method of building
fireproof needs to return. These tiles were laid up and serve for
inner partitions in a house in France (Fig. 7-8), but in California,
they would have to be reinforced against quakes. At one time homes
and factories were built of clay tile.

Figure 7-8

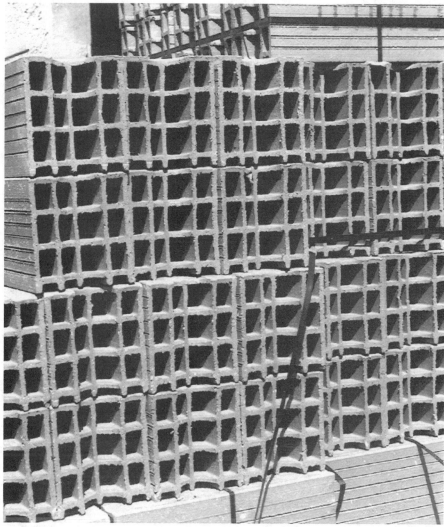

Clay tiles.

Figure 7-9 illustrates a standard way of building inner walls for houses in France. The clay tile will be plastered to match the rest of the house. This partition was built in a 200-year-old dwelling, bringing it up to modern-day standards. Houses in France are built so they don't burn down.

Figure 7-9

Clay tiles in place. These will be plastered to match the rest of the house.

Adobe

ONE of the best building materials for insulating against heat and cold, and at the same time being fireproof, is sun-baked brick, or more commonly called adobe block. Such structures are found mostly in the dry Southwest of the sunshine states.

Adobe is dirt mixed with straw, pressed and sun-dried, so it will hold together when used as building blocks. The quality of adobe is now controlled in many places by building codes. If blocks are to be made by the homebuilder on the job, they must pass a hardness test and resist absorbing a certain amount of water.

Adobe building materials might be defined as earth with 20 percent clay, to which is added four percent asphalt emulsion, a byproduct of the oil industry. (The price of the emulsion fluctuates the same as the price of gasoline, so be sure to check the latest trend when estimating a job.) A few plants manufacture adobe blocks, but because of their weight and bulk, they are rarely shipped any great distance. It is best for the homebuilder to make his block on the job.

Use a plaster mixer to stir the various materials well (not a cement mixer). Then pour the mixed product into molds setting on some type of roofing material, placed on level ground or a concrete slab. If you choose not to use thin roofing material or plastic as a separating material, grease on the slab will make it easier to remove the finish product.

A West-Coast tradition

The padres on the West Coast used adobe extensively in building churches, their rambling sleeping quarters, and the homes of the neophytes. After the buildings were later abandoned, they remained intact until the roofs were destroyed by vandals. Then, the walls gradually crumbled from winter rains, leaving, in some cases, well-defined mounds of dirt, outlines of the defunct buildings (Fig. 8-1).

Today, some of the mission buildings have been restored to show how the native Americans lived under the guidance of the padres. The adobe structures were rebuilt according to California building codes as they pertain to adobe, rendering the building virtually earthquake-proof (Fig. 8-2).

Earthquakes have always been the greatest enemy of adobe. In the 1912 quake, extensive damage was done to all adobe structures the

Figure 8-1

Mission Santa Inez, California, before restoration.

Figure 8-2

Restoration of the Santa Inez Mission by the Civilian Conservation Corps, during the Roosevelt Administration. Rebuilding was required to meet California's newest building codes encompassing earthquakes. California Historical Society

length of California. San Juan Capistrano, where the swallows come each March to build their nests in the remains of the old church, was destroyed by that quake and has never been restored. Adobe is a viable way to build, especially in the dry lands, but only if it is properly reinforced.

Adobe today

The adobe building built according to California standards is a safe shelter, cool in summer as well as impervious to winter cold. Adobe walls are fireproof, reason enough to class the material with fireproof construction.

An adobe building today would be interspersed with concrete columns every 10 feet (3.05 m). Reinforcing is then added, according to engineered plans. The upright columns would be tied in with reinforced bond beams circling the building (Fig. 8-3). This framework could not topple; only individual panels between the reinforced columns would be vulnerable, but even these are unlikely to be damaged.

Post adobe

Post adobe is designed to eliminate the use of reinforcing steel and concrete (Fig. 8-4). Instead, wooden posts (12-x-12 (30.48×30.48 cm) or as specified on the plans) are to be used. The post supports are set approximately 10 feet (3.05 m) apart and bolted to straps embedded in the concrete floor or foundation. T-straps are used to bolt the upright posts to the horizontal beams that will support the roof.

This type of structure has its drawbacks. If backed up against a brush-covered hill or mountain, one is asking for trouble. There is no way to stop all brush fires in the mountainous terrain, so, most likely, the adobe residence will meet with the inevitable fire. The adobe will weather the fire, but the wooden posts or eaves won't. A persistent fire will eat away the eaves and posts on the brush side backed up against the mountain. However, if you cover the eaves and posts with

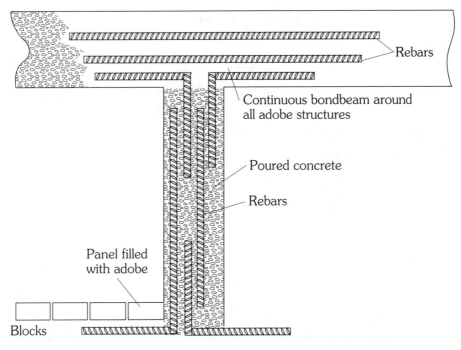

Figure 8-3

An adobe building today is reinforced with concrete columns and bond beams.

stucco and tile the roof, the house is generally able to withstand an extensive fire (Fig. 8-5).

Eaves should be boxed in and plastered, especially if they are lumber (Fig. 8-6). You could use a Milcor plaster ground along the exterior of the concrete bond beams to eliminate a crack from perpetual shaking. Rib lath will span two feet (0.6096 m) and should be used as a plaster base. Use screwable ceiling joists at 16 (40.64 cm) inches OC or metal trusses at two feet OC (0.6096 m). For the interior, wallboard will suffice.

Exterior and interior walls

If your client wants the exterior walls stuccoed, the lather can cover the walls with the proper netting for a stucco job. Stucco directly

Figure 8-4

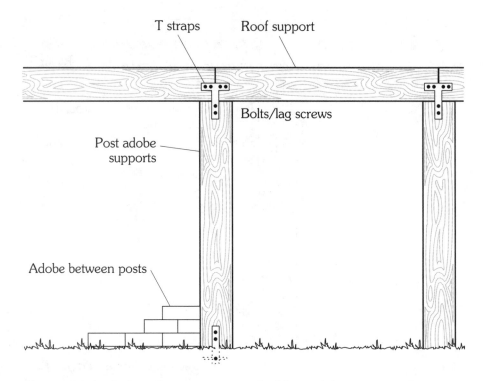

T straps Roof support

Bolts/lag screws

Post adobe supports

Adobe between posts

You might think this is an easier way to build an adobe structure, but it is not recommended. You will be feeding termites, as well as adding wood to a potential fire. If you are going to build adobe, build fireproof, as it should be, with concrete supports instead of wood.

Figure 8-5

If it had been covered with a good roof, more of the mission would have remained.

Figure 8-6

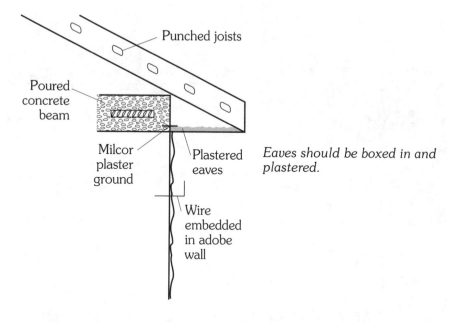

Punched joists

Poured concrete beam

Milcor plaster ground

Plastered eaves

Wire embedded in adobe wall

Eaves should be boxed in and plastered.

over adobe walls will fall off gradually over the next 100 years. You can secure the netting to the walls by pushing a long, L-shaped piece of #9 wire through the adobe or in at least a foot.

If you cover the interior walls with a netting, the same wire can be bent over to secure the netting on the inside walls simultaneously. The covering will hold the plaster to the walls much better than will the adobe alone. If you foresee any difficulty in pushing the wire through the walls, you can install it progressively as you lay up the blocks. If you go to the trouble of putting stucco netting over adobe, you can build the same building in a much wetter climate (Fig. 8-7).

Roofing

Adobe, when used as building blocks, needs to be covered with a wide roof to protect the blocks against inclement weather. When subjected to an appreciable amount of moisture, adobe will gradually crumble over time. For this reason, such blocks are used only above ground and in dry climates (Figs. 8-7, 8-8).

Figure 8-7

Santa Inez, the church and tower restored. Open to the public with demonstrations on how the native Americans lived under the supervision of the padres.

Figure 8-8

Santa Inez, now part of the California State Park system.

To completely fireproof an adobe dwelling, you have to build the roof fireproof. Be sure to read the later chapters that discuss fireproof roofing.

Special caution for dry-land builders

If you, as a homebuilder, are building in the dry lands, you should start with incombustible materials, or you might lose your building before you get a good start. Wind is a contributing factor with fires. A spark in contact with dry brush or grass will start a fire instantly. According to safety regulations, a hose and a barrel of water should be on hand at all times.

One morning while building a fireproof house, a worker was cutting steel joists; a few sparks flew into the grass and started a fire. Another man who was installing joists nearby on the eaves of the garage started to complain of the intense heat so early in the morning. He decided it was going to be a hot day. When he turned around, he noticed the grass on fire. Fortunately, there was a shovel nearby; he started covering the burning grass with dirt. The man who had started the fire never looked up until the first called out, "Bring the hose!"

That could have been an extensive fire, except the neighbor had graded his land, eliminating nearly every blade of grass. Only a few weeds and a native bush along the property line burned.

Fire departments in some desert areas have tried to back-charge anyone starting a fire accidentally or otherwise. I have heard that a welder was billed $26,000 for a brush fire he started.

Steel-stud walls

THOSE in the steel industry can give you many reasons why you should change from wood to steel when framing a house. Few, if any, mention the fireproof nature of the material as one of the best reasons, although they should. They might tell you, for example, that insurance you carry during the process of building, for the first year, would be two-thirds less. Then comprehensive insurance should continue well below that of wood frame (15%)—although I must tell you from experience this is a battle you might have to fight with your insurance carrier.

At one time, the cost of steel studs might have been prohibitive, but this is no longer the case. If you are thinking of building or remodeling, you will find steel studs to be the most economical and the easiest to assemble when compared to other fireproof methods, block or brick. In fact, in many areas, steel studs are even less expensive than lumber.

An education gap really shouldn't be a problem in building with steel studs. If the union is involved, you will find that steel lathing has always belonged to the lathing trades. Steel lathers are sometimes a separate branch from the nail-on crews, but in less-populated areas, some tradespeople do both. Steel studs for inner partitions of multistory buildings is a method that has been used for years, so you should certainly be able to find knowledgeable workers in this trade. They only need to change from tie-on of metal lath with wire, as most work was done at one time, to the screw gun, which is much simpler. Tie-on takes time and is nearly obsolete, but it can still be used for a very small jobs if the screw gun is not readily available.

Construction with steel

Perhaps the simplest way to construct a fireproof wall is to use steel studs, four inches, six inches, or eight inches (5.08, 10.16, or 15.24 cm), depending on the amount of insulation needed in your area either against heat or cold. Twenty-gauge steel studs will suffice for most bearing walls, but your plans might call for a heavier gauge, especially if you are going to use a tile roof weighing 9.2 lbs per square foot (4.40 kg per 0.092 square meter). Inner nonbearing partitions might be 1⅝ inches (4.127 cm), 2½ inches (6.35 cm), 3¼ inches (8.25 cm), or 3⅝ inches (9.189 cm).

Both brick or block, because they are rapid conductors of heat or cold, have to be furred out and insulated in compliance with the newest building codes. If you use the right size steel studs, you can add the proper thickness of insulation with one layer of batting.

As I began writing this book, no steel was required in the footing of a structure incorporating steel studs. Now, in this region, half-inch (1.27 cm) rebars are required at the top and bottom of the trench for bearing walls. The earthquake or hurricane strap is then fastened under the lower horizontal steel. In a wind, the entire slab would have to roll up like a rug, and the strap would still hold the building down. Bolt the track to the foundation accordingly, and perhaps 12 inches (30.48 cm) from the end of each section of track. Since you would not know in advance where this bolt would be inserted in the wet concrete, you might use an extra bolt or two along each outer wall. If you build over two stories, the footing steel framing might be less than for wood framing because the steel system is lighter than the wood counterpart (Fig. 9-1).

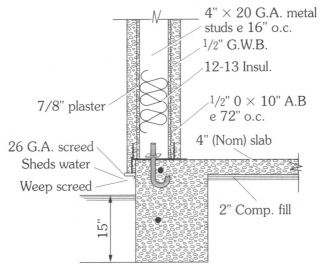

Figure 9-1

4" × 20 G.A. metal studs e 16" o.c.
1/2" G.W.B.
12-13 Insul.
7/8" plaster
1/2" 0 × 10" A.B e 72" o.c.
4" (Nom) slab
26 G.A. screed
Sheds water
Weep screed
2" Comp. fill
15"

Foundation reinforcement.

Everyone knows that steel used as girders will bend and sag in a hot fire much faster than wooden beams. Heavy wooden beams should be plastered, or covered with a layer or two of wallboard to fireproof them. In many instances this is required by the building code. This covering slows the fire and the wooden beams may never be damaged. If the covering over wood is destroyed, the material then feeds the fire, while steel does not.

Insulation

Assuming that wood should be protected by incombustible material, so should steel. There is no reason to use steel framing if you don't intend to follow through and insulate against a potential fire.

Steel won't warp or bend out of shape unless the flames become hot enough to cause the steel to become malleable. In fireproof structures, the contents of the building cause or add to the fire. Generally, by the time the fireproof covering drops from the steel, the fire will have spent itself and the structure will be saved. But if the studs have been covered with a shot-on fireproofing material, backed with fiberglass insulation, this will give double safety to the structure.

In the MGM fire in Las Vegas, where lives were lost, the offices where the fire started were built of 5⅜-inch (13.652 cm) steel studs covered with wallboard. The covering was lost and had to be removed, but the fire stopped there. The studs were saved and re-covered with wallboard. Had the partitions been of wood, the fire would have then had a good start, fed by the wood.

Using steel studs, the job will be extremely level and straight because of the uniformity of the material (Fig. 9-2). Steel studs cut at the

Figure 9-2

Diagonal bracing. Note one screw per stud.

plant will be the same length and the same width. Steel will not shrink in the sun, and the track (or sleeper) will leave your walls without bulges, as so often happens with wood sleepers when they buckle. But if the studs have been covered with a shot-on fireproof material and backed with fiberglass insulation, this will give double safety to the structure.

In a one-hour firewall over wood, after the wall covering goes, the fire has just had a good start, fed by the wood. The one-hour fire rating is designed to allow the fire department time to arrive and extinguish the fire. The building is saved, but it still needs to be partially rebuilt. We need to build so that there is less to burn. Then we can turn our minds to making the contents of the building incombustible.

Steel studs

Steel can be used as either bearing walls (20-gauge) or a lesser gauge for nonbearing walls (25-gauge), depending on the gauge of the material and the width used. With the continued high price of lumber, steel has become cost-effective.

With new equipment, steel studs can be cut as fast or faster than wood. When steel studs are stacked together, they form a small package and two can be cut more easily than one can cut a single 2×6 (5.08×15.24 cm). Then again, you are only cutting fire blocks and odd lengths. The saving comes when studs are shipped the proper length from the factory and no cutting is necessary. A small package of 64 pieces of twenty-gauge track would form a bundle less than two inches high (5.08 cm). The same number of 2-x-4 (5.08×10.16 cm) studding would involve a bundle ten-feet-plus high (3.05 m).

To install studs, place the U-shaped track on the bolts embedded in the concrete slab, either for outer walls or for a partition. Tap the track lightly above the bolt, being careful not to ruin the threads on the bolts. A little coloring from a crayon might help to better transfer the location of the bolt to the track. Drill the holes in the track to fit the foundation bolts. Place the track over the bolts to make sure that everything fits. Place the track on a bench and fasten studs at 16 inches (40.6 cm) OC (Fig. 9-3).

Figure 9-3

Assemble your walls on a bench. A screw gun is a necessity on any steel job.

Fasten the studs to the bottom plate with one screw on each side. Some inspectors might require two screws. Figure 9-4 shows one screw on each side of a steel stud, with 20-gauge sheer straps and a gusset. The gusset might be eliminated in some cases.

The top plate is another length of track; fasten the studs there at 90 degrees. For a header over doors and windows, fasten a second length of track on the underside of the top plate. Then fasten cripples between that and a piece of track over the opening. Fasten in place with screws (6-gauge hex-head or Phillips-head screws). Some welding might be required on multistory buildings and large apartments (Fig. 9-5).

The window will later be screwed to the track and the studs surrounding it. With wood framing, one uses an extra 2×4 (5.08×10.16 cm) next to each opening for reinforcement. With steel, you should use a heavier-gauge stud along the side of each opening instead of an extra 2×4 (5.08×10.16 cm) or cripple as with lumber (Fig. 9-6).

C-studs and joists are manufactured by Cemco of California, which manufactures expanded metal products. They produce galvanized studs from 14-, 16-, and 20-gauge in the following widths:

> 10 inches (25.0 cm) > 3⅝ inches (9.3 cm)

> 8 inches (20.3 cm) > 3½ inches (8.9 cm)

> 6 inches (15.2 cm) > 3¼ inches (8.3 cm)

> 4 inches (10.2 cm) > 2½ inches (6.4 cm)

Figure 9-4

Place one screw on each side of a steel stud.

Flanges are available in 1⅜ inch (34.9 mm) and 1⅝ inch (41.27 mm) with a ⁹⁄₁₆-inch (14.29 mm) return for extra stiffness. Studs are available punched or unpunched. The track is bent slightly inward so the stud, when placed within the track, is gripped and held in place for a screw to be applied, one on each side or according to the inspector.

Figure 9-5

Two-story or multistoried buildings might require studs to be welded instead of screwed.

You should be able to find steel companies in your area with an equal product. Each steel company producing studs will furnish the dealer with booklets of the weight, gauge, and capabilities of their product. These booklets are important for engineers when producing calculations for different jobs, and they are useful for the homebuilder when estimating a job. Usually, the company will photocopy the pages you need or give you an entire book if they have enough (Fig. 9-8).

Figure 9-6

This house uses a 2×6 or 8 (5.08×15.24 or 20.32 cm) as a top plate instead of steel.

When you have assembled a section of wall, lift it back onto the bolts in the slab. Have a 2×4 (5.08×10.16 cm) ready to brace the section of wall until it has been properly supported and bolted down. Bolt the section to the floor using washers, if required, to cover a greater area of metal so the wall will never separate from the track, even in high wind (Fig. 9-9)

Figure 9-7

Use heavier studs for multistoried buildings. Note the holes punched in the studs to accommodate wiring and plumbing.

Figure 9-8

This public building used steel studs as framing and has the headers over the doors and windows welded.

Earthquake straps

The newest requirements in California construction is the sheer strap. Where a large area of the exterior wall is not bolted to the foundation, such as an 18-foot (5.49 m) door opening, in wood, a 4×4 (10.16×10.16 cm) is used next to the opening. An earthquake

Figure 9-9

Here you can see some wood still evident in the framing. I say leave the forests for the owls.

strap, or a 2-×-¼-inch (5.08 cm×6.35mm) strap is embedded into the foundation pour and then screwed or bolted to the steel framing. This newest innovation keeps the building from jumping off the foundation from the initial jolt of an earthquake or from a severe wind. Several types are available on the market today (Fig. 9-10).

Figure 9-10

Earthquake strap.

In Fig. 9-11, the earthquake strap is screwed to the corner stud. Bolt holes are provided for the use of this method. Sheer straps were bolted to a 20-gauge gusset, and, as you can see, an excessive number of screws were used to assemble the track and gusset. The gauge of the gusset should be evident on your plans. Be sure you work these details out with the engineer as the plans are drawn, or you might have to work them out on the job.

Fastening

Also available on the market today is an explosive gun for pins with the washers already intact. You simply fasten the track to the floor with the gun. This simplifies matters by eliminating bolts, but go by the plan drawn for your area. Bolts are still required for the outer walls, especially in hurricane- and earthquake-prone states.

Figure 9-11

Earthquake strap screwed to the corner stud.

The metal wall section is so light that one man can handle it if necessary. Screws used to assemble the wall are supposed to be several times cheaper than nails. You might not find this so, but you will use fewer screws than you would nails. If necessary, snap a chalkline to properly align inner walls. With a metal cutting blade in a Skilsaw, you can cut the door openings in the track after it is securely in place.

Figure 9-12

Twenty-gauge punched, galvanized studs. The darker material is 14 gauge.

Always wear goggles when cutting any metal with power tools. Little chips of steel could embed themselves in your eye (Fig. 9-12)

For inner partitions, to fasten the track to the slab, use the power gun with a pin, washer, and bullet that will penetrate the concrete. If the washer is separate, place the gun against the washer on the track, and pull the trigger (Fig. 9-13). The nail or pin will go through the track and embed itself in the concrete. If you try pinning the track to an outer wall, don't pin the track too close to the edge or you might break off a chunk of concrete.

If concrete, brick, or block walls are used for outer walls, provide bolts to fasten the first stud, or you can shoot pins into the block wall with power gun. Be careful: Too powerful a shell could send the pin through the block and endanger someone on the other side of the wall. For proper results, the power of the shell might have to be determined by trial and error. Shells of various explosive powers are

Figure 9-13

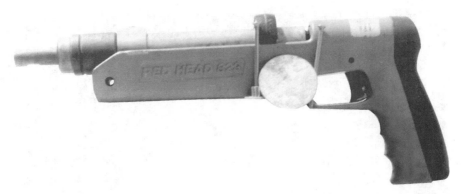

Power gun for fastening track of inner walls to the concrete slab. Note the round washer along the side, for pins that don't have washers provided.

available. Buy pins and shells at the material yard where you bought the studs.

Anyone proposing to use a power gun should learn to operate it safely. In some states, certification is required for operation of the gun. The store where you bought the gun should provide sufficient lessons on how to use this sophisticated tool. A gun should have a safety device so it will not fire without pressure applied against the object to be pinned down. In any event, don't point the gun at anyone in jest or otherwise.

Once the track for an inner partition is in place, with studs bolted to the outer walls, you can place in place your top track, or *plate*, as it is called in wood framing. For short runs, fasten your first stud with screws, on the top and bottom of both sides. For longer or even shorter runs, you might assemble the section on a bench as you did the outer walls (Fig. 9-14).

If your wall is a little shaky, use 1½-inch (3.8 cm) metal straps for bracing diagonally to stabilize the wall. Secure the straps in place with screws. When the electrical is installed, thinwall is better than Romex, which might rub on the metal and, over the years, wear a hole into the fabric, exposing a live wire. Pull electrical wiring through thinwall tubing so you can add or subtract wires without

Figure 9-14

This all-steel-stud building for the Victor Valley Junior College district is scaffolded and ready for lathing. The trend is for all public buildings to be built fireproof. Now it's up to the homebuilders to follow through with residential building.

damaging the walls (Fig. 9-15). The tubing can then be wired to the studs to stabilize the wall.

Wall covering

When you have installed all the studs, put the roof on, and had everything inspected, fasten your wall covering with self-tapping screws that penetrate the studs (#6-gauge, 1-inch (2.54 cm) flat, Phillips head, or equal). When one side of a wall is covered, the metal should be insulated. In addition, you should cover the studs with some type of plaster or shot-on material to make them as impervious to fire

Figure 9-15

Steel studs and ceiling joists come with holes punched in them to facilitate the installation of wiring.

as possible. Zonolite, or its equal, comes in large, lightweight bags and is easily applied with an acoustic gun or by hand with a hawk and trowel. Mix the material in a new trash can with a hoe, and apply by hand for a small job or shoot the area with your own or a rented acoustic gun for a large job.

The rental yard should be able to give you an easy lesson on the application of the material with an acoustic gun. If you learn how to apply the material evenly, you could later shoot your own acoustic ceilings.

Scrape the excess material from the surface of the studs while it is wet so those walls can be lathed with ease at a later time (Fig. 9-16).

Figure 9-16

Two-story steel framing.

You easily can determine how much material you want to shoot over studs. On nonbearing inner walls of lesser thicknesses (two inches/5.08cm), you won't be insulating, so you might want to apply a generous coat of fireproofing material over the steel. Or you might want to install the insulation along the outside walls first and spray over that, filling the wall to the depth of the studs.

You can attach a device to a small drill to allow the use of self-tapping screws. An automatic screw drill might cost $100 or more, but it is worth the money when any appreciable amount of work is to be done with screwable studs.

One last caution . . .

Do not underestimate the importance of engineered plans for proper reinforcement. Figure 9-17 shows a building that failed to withstand an earthquake. The pier that supported the I-beam on the right failed. The heavy I-beam then swung back and forth like a gate, wiping out all the steel studs. I have never before seen a properly engineered building wiped out by an earthquake. On the other hand, I have taken numerous photos of bootlegged programs that were destroyed, where no reinforcing steel or concrete was evident.

Figure 9-17

This building failed to withstand an earthquake.

10

Steel joists
and trusses

UP until now, you have learned to build the walls of your fireproof house from one of four different versions listed in this book. The most neglected part of any fireproof construction, however, is roof framing. Then, nearly all homebuilders turn to a product with which they are familiar—lumber.

Advantages of steel roof trusses

In trying to find a material dealer stressing steel roof trusses over wood, I came up with only one company that bragged about steel. They tell the advantage of their product over wood as well as C-studs, now a common product on the market.

Western Metal Lath and Steel Framing, states: The strength per pound is maximized with their system, due to the flange and rib reinforcing. Their TC member has a *Q factor* (an engineering factor that measures compressive strength) of 1.0. This is supposed to be 66% stronger than common C-studs.

Labor is saved by using Western Metal concentrically loaded hat-shaped truss material. Their material does not require miter cuts, the webs slide snugly inside the top or bottom chord. Gusset plates are not usually required since the members can be fastened from the side through the top or bottom chord into the web member. Conservative estimates show that this material may require 33% to 50% of the screws and about 33% of the labor to assemble as a similar configured C-stud truss.

Western Metal claims that their trusses are lighter and easier to use. Worker fatigue is reduced due to the light weight of their trusses when compared to either wood or other trusses formed from steel studs. In fact, the company maintains that the entire building can be designed lighter (and at less cost) when the weight of the roof system is reduced. A lighter roof does not require as expensive a support as does a heavier roof.

Western Metal trusses can be spliced, overlapped to save waste. Waste is nearly eliminated, since top and bottom cord splices together. This can be a savings of material. Portions of the truss requiring added strength are stiffened, allowing for material efficiency. Rather than throw a scrap of material away, a greater lap may be retained.

Western Metal's truss material is made of galvanized steel for corrosion resistance, which is a general practice now. Since steel does not support combustion, any brand would be equal in that regard. All steel framing is immune to termite and insect infestation.

Steel trusses have greater wind resistance. Because the material is screwed together instead of nailed together, it can better resist the long term and temporary effects of high winds. Trusses can be engineered to withstand whatever wind load is necessary. Western Metal has tested their product by simulating an extreme wind induced uplift condition during the design process. The result is proof of the resistance of their trusses to extreme conditions.

Steel companies have become concerned with the depletion of our forests products. Saving our forests can best be done with the use of steel in place of lumber. At a Sears store, I watched two men remove the plywood from a steel stud wall. Then they unscrewed the studs, and the room was back to its original condition in minutes. A small amount of pointing up of the walls was necessary, then painting, and the room was ready for a different display. They simply stored the studs and wall covering to form another room at a later time. You could hardly do that with lumber.

Where to buy steel trusses

Many companies fabricate a simple steel truss for the gable house. They are assembled at the factory and hauled to the job (Fig. 10-1). The truss made from metal is much lighter than trusses made of lumber. One worker can handle a single metal truss while loading and unloading and maneuvering it onto the roof. When the trusses are installed you have the ceiling joists in place at the same time.

Figure 10-1

Trusses assembled at the factory and delivered to the job by Western Metal, specializing in lath and steel framing systems.

You can also form your own trusses on the job by installing ceiling joists first, then the rafters of metal, and the struts or braces. On the West Coast, however, these have to be engineered. The calculation sheet comes in a package with the trusses when delivered to the job assembled or unassembled. The building department will want two copies of the engineering sheets with the plans for checking.

Engineered plans

Each type of truss has to be engineered. If you have a gable house, you can get by with one type for the house and one for the garage attached or not attached. Keep in mind that an engineer may want $100 dollars for each type of truss he engineers.

The company where I proposed to buy my material gave me an estimate on the steel trusses, a kit to be assembled on the job. The figure included the cost of the studs and the material for the walls. They wanted $120 for engineering the two types of trusses. When I sent them a check for the engineering, after ten days they informed me that they didn't have an in-house engineer, but I could send the prints to the engineer they recommended and he would proceed.

I mailed a blueprint to a different company. They promised to send me an estimate on the engineering and on the cost of the material for the walls and trusses. The second company wanted an extra $4,000 to furnish the steel trusses and wall material. I contacted my engineer and soon found that he doesn't engineer steel trusses or steel stud buildings.

The point of all this is that I should have had the engineer who sent me calculations on the ⅜-inch (9.52 mm) rebars used in the solid partitions of the house. He did the engineering of the solid wall partitions in an hour. I thought it would have been best to have had him engineer the walls and the trusses as well as the two-inch (5.08 cm) partitions. I hadn't built a house in some time so I had to learn my way around. I supposed the company furnishing the material would be the most logical one with whom to deal for engineering calculations, but as you can see, that's not always the case. Even a company that wants more business doesn't necessarily have the time to properly serve the customers they have.

Steel truss construction

With trusses and their overall span of the building, which we in California are now forced to use, one can eliminate all the footings for inner partitions (Fig. 10-2). The figure shows a 38-foot (11.6 m) span, with the porch overhang cantilevered eight feet (2.43 m) both front and back. This might be a saving that hadn't previously occurred to you.

At one time, I built a house with a high gable without trusses. (This was before trusses were required.) See Fig. 10-3. The inspector

Figure 10-2

C-studs span this building.

Figure 10-3

Note the pile of lumber to form a truss compared to the small amount when using C-studs.

turned down the framing because the roof would not have been adequately supported and, over time, it would have had a sway. I had paid the carpenters who did the work, so they refused to return and bring the job up to the inspector's requirements. I had to hire a second carpenter, to whom I gave the red tag that stated the faults of the workmanship, and I directed him to repair the framing so it would pass inspection. He ran 2×4s (5.08×10.16 cm) in all directions to better support the roof. Each brace had to set on a bearing partition, which in turn was supported by an adequate foundation. Now, trusses are a requirement in the High Desert area.

It used to be best to avoid long spans, resting the joists on bearing walls where possible. That is no longer necessary with the trusses all spanning the entire house. Most trusses built in the factory are meant to span two feet (60.9 cm) OC. This is fine now, since building departments in many places allow the attachment of wall board over trusses at two feet (60.9 cm) OC, but not over ceiling joist alone. Ceiling joists alone are still required to be placed at 16 inches (40.64 mm) OC for the application of wallboard.

Rib lath also will span two feet (60.9 cm) but has to be plastered three coats. You might get by with only two coats in factories or firewalls. My own fireproof house has metal lath over steel joists at 16 inches (40.6 cm) OC. In Fig. 10-4, the track for the ceiling joist was embedded in the poured concrete column, surrounded by four pieces of half-inch (1.27 cm) rebar. The ceilings were scratched and browned and the third coat acoustic. School buildings and many public buildings used to be designed for metal lath and three coats of plaster. This is the best way to go for a class-A job. Some will say that plastering has gone out of style, but styles come back when there is a need, and this might be the case when one aspires to build fireproof.

Figure 10-5 shows ceiling joists that span the room and are cantilevered for the eight-foot (2.43 m) porch, both front and back. Lightweight aggregate was used for inside plastering of all metal lath ceilings.

Once the trusses are in place, bolt them to the reinforcing for that purpose, or use the metal bracket that screws into the plate. The

Figure 10-4

Odd pieces of C-joists and bracing will be removed when the column is unveiled.

Figure 10-5

Ceiling joists.

truss is screwed to that. You will have to brace the trusses temporarily so they won't sway. If the end gable is built of block or brick, it will be easy to bolt that truss to the wall. In any case, after all the trusses are in place, block them solid at least along the ridge, or as the plans specify (Figs. 10-6, 10-7).

Steel trusses are formed on saw horses (Fig. 10-8). The trusses are then blocked along the ridge. The trusses in this instance were 16 inches (40.64 mm) OC and served as ceiling joists for the attachment of wallboard or lathing material.

When homebuilders become more proficient, they will no longer need to use precut material. Instead, you can assemble each truss in the simple jig you build yourself. The structure shown in Fig. 9-2 was built by one venturesome man, the owner, who assembled the trusses in a jig and maneuvered them into place. The precut kit of C-joists was furnished by the supplier.

Figure 10-6

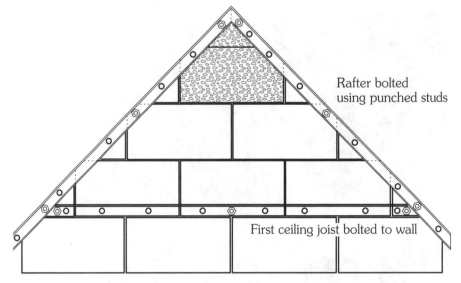

Rafter bolted
using punched studs

First ceiling joist bolted to wall

The gable end of a block house. The first ceiling joist or rafter need not form a truss; rather, it can be fastened to bolts embedded in the concrete poured around the anchor end of the bolts.

Figure 10-7

Steel trusses or rafters will be bolted to the gable end of the block building. (This one is in Ireland where, because there is little lumber, they build nearly fireproof as a matter of course.) Note the airspace between the layers of block. No steel was added to strengthen the building. The island sets on solid rock and doesn't shake.

Figure 10-8

Note the temporary framing to support the first joist.

Preparing for the roof

Different steel companies show in their brochures how to attach plywood to roof trusses as a surface for applying roofing materials. Half-inch (1.27 cm) exterior-grade plywood is then placed on the roof with a forklift where it is attached to the metal joists with a screw gun (Fig. 10-9). A fireproof sheathing will someday be fabricated to replace the combustible. To help negate the fire hazard, you might attach a layer of half-inch (1.27 cm) wallboard over the entire roof after the application of plywood. Over this, apply various layers of felt, using long enough nails to penetrate the wallboard as well as the plywood. Finish the roof according to your print. In this way, you will comply with the current building codes.

Figure 10-9

When you do not buy precut truss material, you will need a metal-cutting blade for your saw. Partially used cutting blade hanging on metal lath. Skilsaw

The material yards stock only galvanized studs and joists. The heavier joists and studs, when special-ordered, are not always galvanized. Be sure to check; you will have to comply with codes that may differ from state to state. Heavier joists can be galvanized on request, for a fee.

In Fig. 10-10, two 2-×-6 (5.08×15.24 cm) plates were used around the top of the building. This was really unnecessary, as the top track alone would have sufficed. The dark material shown in the figure is not galvanized, but painted. The builder of this project sometimes used a heavier-gauge stud, eliminating the use of two lighter-gauge studs. You can sometimes use two 20-gauge galvanized joists back-to-back, rather than ordering galvanized joists of a heavier gauge, say #14.

One final note: Keep in mind that if you plan to cover the roof with tile, your building will have to be designed to carry the extra load.

Figure 10-10

The two 2-×-6 (5.08×15.24 cm) plates are unnecessary in this case. The top track would have sufficed. Note one screw at the top and bottom of the gable. One screw sufficed here, but you should check with your own building inspector. He or she might have different thoughts on the number of screws.

11

Asphalt shingles and rolled roofing

WHEN building fireproof, sometimes you can be forced to deviate from the ultimate, entirely fireproof roof. The building department might tell you that they want a standard roof or they will not give you a permit to build the house. So you turn to something less fireproof, particularly when your plans were not designed for anything as heavy as tile or concrete.

Shingles are the worst kind of building material, mainly because they catch fire so readily. Since shakes are thicker than shingles, there is more to burn, so they burn longer and hotter. The thatched roof is even more vulnerable to fire than shingles. Such roofs are found where it rains often enough that the roof covering is generally damp, making it almost impervious to fire from the exterior. But because the thatched roof is not plentiful today, we can bypass that phase of roofing.

Just as the thatched roof has become obsolete, so do shingles and shakes need to be gradually phased out, with something just as eye-appealing, but fireproof at the same time. Many people think of shingles as beautiful, but that's a matter of opinion. It's difficult to conceive of shingles being beautiful when you know they can eventually become tinder and easily go up in smoke.

In many European countries, houses are so close together that one can jump from roof to roof. Since the roofs are of tile, no fire starts from the roof but is generally limited to the contents of the building.

In the United States and Canada, forest-product companies must have a powerful lobby to keep shingles on the roof of houses when there are so many superior materials as far as fire safety is concerned. If we would have had a similar, powerful lobby at the time of the changeover from the horse to the car, we would still be sitting in a buggy, looking at the rump of a horse trotting down the road every morning as we traveled to our jobs. We have advanced to the car, which shows we can make progress. Why can't we advance in the construction business and make our residence a safer place to live?

With shakes at $100 a square, plus labor, and plastic shingles simulating shakes not far behind, we should turn to the all-stucco roof, wherever appropriate. If built right, a stucco roof would not have to be replaced for 100 years. Unfortunately, stucco roofs should never be built where moisture penetrating the material might freeze. Check with your building department to ascertain what can be built in your area.

In Europe, they have for several centuries built houses with walls and beams strong enough support three tile roofs. Recently, on the house where my great-grandfather lived, a new tile roof was applied—only the third in 200 years.

Making the best of asphalt

Asphalt shingles are similar to wood shingles, slices of roofing nailed together to shed water. They come in various colors, some not too dissimilar to wood, especially before wood darkens with age.

When fabricated, asphalt shingles and rolled roofing, 70-lb. (31.75 kilograms) felt and 90-lb. (40.82 kilograms) felt are sprayed with a coating of hot asphalt, followed immediately by a layer of colored rock, called mineral. The mineral renders the surface of the different materials somewhat impervious to fire from flying sparks or small burning objects. The mineral surface also protects the asphalt product from the hot sun, causing the material to wear longer. This type of roof covering had been applied to one of the houses spared in the recent Anaheim conflagration. The mineral prevented the roofing from kindling less readily, so the house was saved.

But once the asphalt melts sufficiently to catch fire, it will readily go up in smoke.

The mineral-coated felts and asphalt shingles are virtually the same. The shingles are used on steep slopes where they are displayed for beauty. The water cascades off the roof without entering the building. Rolled roofing is used on commercial buildings and flat roofs where the material is not readily seen. If the roof is mopped solid, water is prevented from entering the building, even where a lake might form on the roof.

Gravel stops

You can buy a gravel stop wherever you buy the rest of your roofing materials. It is installed before the first sheet of base coat is applied (Fig. 11-1). At one time, one never used a gravel stop except when applying rock over the last layer of roofing, but now many roofing contractors attach a gravel stop before they roll out any roofing.

Figure 11-1

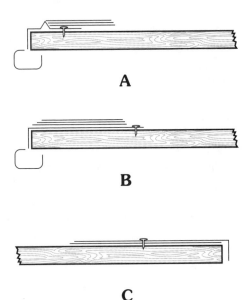

A

B

C

(A) Gravel stop, needed to keep the gravel from blowing off the roof. (B) L-flashing drains the roof better. (C) Upper part of the roof. L-flashing is applied over the final cap sheet, preventing water from entering the roof better than does the gravel stop, especially when the roof becomes old and shrinkage occurs around the metal guard.

I find a gravel stop has a tendency to retain water and causes a lake to form at the lip of the roof on rather flat areas. To avoid this, brush a little lap cement onto the gravel stop. This will allow the leakage to drain off instead of backing under the first layer of felt.

If you insist on using a gravel stop, the best way might be to install the first layer of felt, then the gravel stop. In this way if water backs up from the gravel stop, there is no way for it to enter the eaves, or house, as it drains off. Many times I have used an L-type flashing, applied before the first layer of roofing has been installed. The lip diverts the water into a gutter and 2-x-10-inch (50.8-x–254-mm) L-flashing prevents any water from seeping backward onto the sheathing.

Applying shingles

First, cover the roof with a layer of 15-lb. (6.61 kg) felt, lapping a few inches. Don't hurry and cover the entire roof; the wind might blow the material away. Apply the felt progressively as you shingle. Lay the first row of shingles backward over the 15-lb. (6.61 kg) felt, with the slots upward, mineral up. Lay the second row with the slots down, directly over the first row. The slots are only for looks. The shingles actually would be better off without them. Water cascades over the jump off, causing the shingles to wear between the slots much faster than the rest of the shingle.

As you continue to install shingles, leave the slots in a straight line. With the rows even, you will attain a beautiful, symmetrical roof. There is a place noted for each nail so it won't be exposed to the weather, but rather covered by the next row of shingles. Follow the directions on the package, overlapping the joints and you will have no leaks.

Asphalt shingles are much cheaper than wood shingles. They might not last quite as long, but will weather more of a fire from burning debris flying through the air.

I once applied stucco over a standard roof covered with one-inch (25.4 mm) chicken wire. Later, an electrical storm broke the overhead power line. A hot wire fell onto the roof and burned a small amount of asphalt. The fire died down when it encountered the stucco, otherwise I would have lost a roof. The walls were built of cement block (Fig. 11-2).

If you are building up a flat or almost-flat roof, you will have to start at the lower edge with the first layer of felt.

Your plans might specify two layers of 30-lb. (13.63 kg) felt or three layers of 15-lb. (6.61 kg) felt. Nail the material down according to current area codes.

Figure 11-2

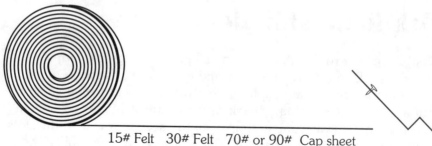

15# Felt   30# Felt   70# or 90#   Cap sheet

We are sometimes forced to turn to asphalt products.

Mopping

Due to earthquakes in California, a roof mopped tight to the building by either sprinkle mopping or strip mopping will shake and tear the roofing material when the roof moves suddenly and excessively during a quake. Similar problems are found with wind. Because of these problems, mopping directly onto the sheathing should be discontinued. The base coat, nailed but not mopped, floats with any movement of the roof and will probably escape running crack damage, at least until the material has lost its elasticity and becomes brittle with age. Then the material can pop when shaken and cause a running crack.

A friend of mine with three unemployed sons, found I had roofed a number of buildings. He thought he would roof one of his own. I gave him and his sons all the advice I deemed necessary and went home. They were too slow in using the tar as it melted in the kettle. Instead of turning the heat down, they allowed the material to overheat until it reached the flash point. The lid on the tar kettle blew up and dropped back down. The tar that spilled on the kettle caught fire; before they were able to contain the fire, both tires on the rented machine had burned. When using hot mop, always keep the hose or a barrel of water handy. (The boys didn't give up, but instead overcame their difficulties and are now licensed contractors in the roofing business.)

Today, all up-to-date tar kettles have heat gauges. As long as you do not allow the heat to exceed the recommended temperature on the packaged asphalt, you will have no trouble with fires.

Flat roofs

Flat roofs or almost-flat roofs give the most trouble, but they will not leak if properly mopped. On flat roofs of built-up rolled roofing, when the material loses its resiliency, running cracks sometime develop from heavy jolts. The use of fiberglass-entrained material helps to prevent such cracks.

When installing roofing material on an extensive roof, you should cut the material, allow a lap and start over every 25 feet (7.62 m). The next row, break joint and continue the same way with alternating joints. The double, reinforced joints will help to prevent cracks or breaks from the shaking of the building.

The first layer of felt serves as a vapor barrier, the second and following layers of felt are mopped solid with asphalt. This is especially important on a flat or almost-flat roof. For two layers of felt, mop for a full sheet and move up half a sheet; for three layers, mop for a full sheet and move up one-third of a sheet, applying the felt over the newly applied asphalt.

Precautions for wind

Nail the top edge of each unmopped base sheet as needed as it is applied. In windy areas, a special nail, BRT A/T square cap 1½-inch (38.1 mm) might be required, or an equal where those nails are not available.

I saw one case, on a 10,000-square-foot (92.90 sq. meters) factory, where the roof blew off a few months after the building was finished. The first layer of paper had not been properly nailed. Inspectors don't catch everything. Roofing contractors should be responsible and follow through according to plans. Any error on the part of a subcontractor reflects on the homebuilder. I presume the nailing schedule on the plans was adequate if it had been followed.

As a roof shakes, the nails finally wear a larger hole in the roofing. Try and avoid nailing excessively where water might stand for a time before draining away. This will prevent water from creeping in around nails when the roof becomes older. Bear in mind, in windy regions, the roofing contractor doesn't want to lose a roof from lack of nails. Apply all material according to area codes.

Adding rock

After the entire roof has been laid, mop it again with asphalt in small squares. Before the tar hardens or immediately (whichever is quickest), sprinkle small rock, #2 or #4, over the hot surface. White rock is best, as it absorbs the least heat. If the tar hardens before the rock comes in contact with it, the rock will not stick to serve its entire purpose.

Without rock embedded into the tar, roofing material will lose oils more readily and the felts deteriorate more rapidly. Even the few rock that do not stick to the tar reflect the sun to keep the roof from getting too hot in summer, which is what rock are meant to do.

I find a rock roof stands up well, but I no longer use this type of roof because I live in California. When a quake shakes and finally damages the roof from running cracks—either parallel to the paper rolled out or perpendicular to the installed roofing—it is sometimes difficult to find the crack.

You can repair leaks temporarily by scraping the gravel away and mopping the strips with some kind of roofing material over the break. Then apply a strip of mineral-covered roofing.

Using cap sheet

With a cap sheet, running cracks are easier to find and repair. Over one or two layers of base sheet, nailed, you might apply two layers of 11-lb. (5.06 kg) fiberglass roofing material, mopped between layers. The fiberglass helps to keep the material from tearing easily. Next apply a fiberglass cap sheet, 70 lb. (31.75 kg) or 90 lb. (40.82 kg), in your choice of color, hot-mopped.

Buying materials

It's a good idea to buy your asphalt locally. In hot areas, the product is designed to withstand the intense heat without melting and running off the building. The same product sold in cooler areas would not suit desert conditions.

Various kinds of roofing materials are available. Discuss your design and application needs with your dealer. While nationally known products are readily available, it might better suit your needs to buy from local plants, whose products never get beyond adjoining counties. In almost every case, these products are just as good as the more expensive, nationally advertised kind (Figs. 11-3, 11-4).

Figure 11-3

Asphalt by the 100-pound (45.4 kg) keg.

Figure 11-4

The mineral on all products will keep small fires from catching hold.

If you are remodeling, a solicitous dealer will furnish you with mineral to match an existing roof. Use it around the vents and joints, where you have applied a fibered roof coating. Tamp into the soft material. This beautiful match will satisfy both the inspector and the lending institution (Fig. 11-5).

Figure 11-5

Sometimes a dealer will provide you with a mineral to match existing roof material.

12

Metal roofs

IF you follow the construction business, you no doubt know that there are fire zones in the inner cities that require the use of fireproof building materials. I had the opportunity to work on such a project at an exclusive night club in Beverly Hills where membership is $1,000.

On this job, I attached rib lath, 3.4-lb. (1.4 kilograms), over wood ceiling joists on a nearly flat roof. No paper was required under the lath, but we fastened #9 wire to the joists every six inches (15,24 cm) as tie wire. After the lath was nailed down, you could walk on the roof. Over this, I applied a cement coat with pea gravel to prevent the material from falling through the webbing. A second coat of plastic cement and sand was applied over this, which was darbied (straightened out) and floated.

This completed the plastering for the exterior. The sheet metal contractor installed a copper roof to fulfill the fireproof requirements of the city. That roof would never catch fire and would outlast any owner. In most residential circumstances, though, one has to re-roof regularly with wood shingles or asphalt products. A copper roof may be somewhat expensive, but so are asphalt products. With a copper roof, your client will never have to roof again. You can certainly look into the feasibility of metal fireproof roofing for residential buildings.

The metal roof, once used only in the industrial field, is moving into the commercial. It often forms a fancy porch-like front, a protection against the sun and rain in shopping centers, allowing people to shop and stay dry in any weather conditions (Fig. 12-1). A few of these innovations have moved into the residential area.

In the tropical areas, you see metal roofs. In Hawaii, they use plain corrugated iron. Now, we have on the market the metal roof of different gauges, 22 and 24 etc. Roofs with seams one-and-a-half inches high present a clear, bold-ribbed design of beauty. There is no chance for water to enter the house since the seams allow for a sealant, either built in at the factory or applied by the installer. The seams are then tightened with screws and fitted with neoprene washers that compress the panels together to exclude all moisture. The screws at the same time penetrate the roof structure or built-up cross members to secure the ribbed panels. No water ever enters the

Figure 12-1

Metal roof in a commercial building.

joints or house, even in a driving rain as found in the wind-prone area. The material is locked together so it would seem that a single metal slab is protecting your building.

If you, as a homebuilder, are working in a forested area or any other area of high fire potential, you should consider this type of everlasting roof for your clients. Metal roofs have the durability that architects demand and are manufactured for slopes of 2:12 and greater. Some factories prepunch all holes for assembly with expanded slots so that, after installation, with expansion from heat or cold, the metal is not torn at the slots where screws penetrate.

Most material comes hot-dipped galvanized steel or hot-dipped aluminum-zinc. This type of roof paneling is used mostly where it is not seen. In some store fronts you can see them, but most metal is covered with a special resin of appropriate color to match the architectural design. The preparation lasts so well it might be termed an everlasting paneling. This type of roof has to be planned in the design stage, since we know that paint will not stick to galvanizing.

Metal roof construction

These days, in compliance with most inner-city fire zone requirements, metal ceiling joists, perhaps C-joists, would be used on the night club I described. With the extra kink, they give the most strength for your money. After lathing such a ceiling—metal lath, wallboard, whatever—drop the insulation in place from above. You could then shoot a coat of fireproofing over the remaining surface of the joints, filling in all voids where the insulation doesn't fit tightly.

Recently, in compliance with a fire zone, I installed 10-inch (25.4 cm) C-studs for a nearly flat roof with ceiling joists as mentioned above (Fig. 12-2). Joists were doubled around the opening for the skylight. It's difficult to tell from the photo, but the ceiling joists slope, and they were leveled with hangar wires, then lathed with 3.4 rib lath and plastered.

Figure 12-2

C-studs used as ceiling joists.

To level the ceiling, I shot black metal to the wall on the lower end of the rafters. Then, on the high end of the joists/rafters, I secured black metal in place by the same method, but level with the first strip. The rest of the black metal strips were then wired to the rafters at four feet (1.22 m) OC, using #9 soft wire, and level with the strips attached to the wall, as one would prepare for a dropped acoustic ceiling. I wired furring strips flat, at 16 inches (40.64 cm) OC to the flat metal installed on edge, with #9 soft wire (Fig. 12-3).

Figure 12-3

Hat furring and other metal furring available at the material yard.

With the drop ceiling leveled, I attached rib lath to the furring strips with self-tapping screws. I made sure that each screw was placed in the heavy metal, not in the expanded area where it might readily pull out in some type of a disaster. Talk with the inspector to make sure you are using the correct screws (Fig. 12-4, 12-5).

Figure 12-4

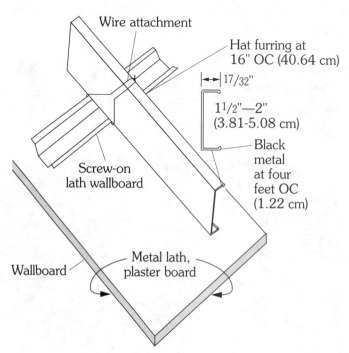

Detail of furring assembly in place of ceiling joists.

With the lath in place, I dropped the insulation (R-29 with no paper backing) in place between the rafters. This is really the best way to handle such a situation—since insulation cannot be attached to the steel with a staple gun as with lumber. With the permission of the inspector, who didn't want the insulation compressed when I shot the fireproofing over the steel joists, I used Zonolite, the only fireproofing available in the area.

I shot a heavy coat of fireproofing over the metal joists on both sides. The rebound landed on the insulating material. This filled all voids

Figure 12-5

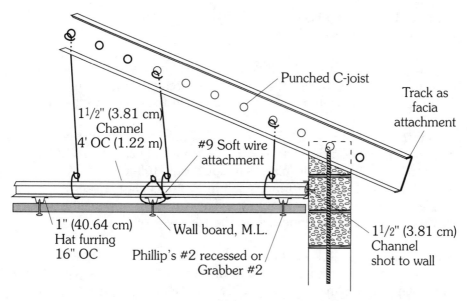

This view of a dropped ceiling would also be attached to the ceiling joists.

between the joists and the insulation and left a solid cover. The inspector was happy. I was encouraged to finish the roof before it rained. The fireproofing of the joists might not be required, but you should apply it because steel will bend out of shape in a hot fire. Nowadays, all steel of multistoried structures is fireproofed by this method before installing any inner or outer wall covering.

To the metal lath I applied two coats of hardwall, roded the second coat, leveled and floated it for an application of acoustic material. There has never been a crack in the 1,200-square-foot (111.48 square meters) living room ceiling in the house I once plastered according to these regulations. The ceiling floats and seems to withstand the shocks of nature with no ill effects. The walls of the house were fireproof. If the contents of the house ever burns, only the plastering would be affected—by heat, then water damage, and that could be reworked.

At the same time, shoot the underside of the gable ceiling if you have room to work there. That will not only fireproof the plywood but also the ceiling joists and the rafters; it will also insulate against heat, and, or cold.

The inspector might require that all electrical boxes in the attic be covered with metal lath so they are not entirely covered with fireproofing. Then the metal can be removed for access to the boxes.

One inspector in the area gave me a hard time because I lathed the ceiling with ML and dropped in the insulation. He said the roof had to be on before I could insulate. He could only think wood. With wood joists, you easily fasten insulation to the ceiling joists with a staple gun, but not with metal joists. There was no appreciable attic, so the work had to be done before the roof was on.

The original inspector went on vacation, so I called and asked for another inspector, who agreed with me. From then on I requested the more sympathetic man to inspect the job. Metal joists and studs were few at the time and the new inspector appreciated the innovations; he was simply fascinated with the job. One day he brought his boss with him to show him what was happening in the way of fireproof construction. The inspector confided in me that he was being moved to a territory closer to where he lived. His only regret was that he would not see the house in its completed stage.

13

Cement slurry roofs

HOME fires should not be taken lightly as though nothing can be done about a potential catastrophe. Thousands of houses burn every year in the United States. The homebuilder or the roofing contractor can make an effort to stop some of those fires from happening by surveying the needs of your own roof, then following through.

People can't afford to start over and build an all fireproof house, but they can upgrade what they have. Many homes burn to the ground every year starting with the roof. Nearly all houses have combustible roofs; you can easily and for little money or effort render some of those roofs nearly impregnable to fire.

A few companies do nothing but spray roofs with a slurry of cement to insulate houses, rendering them warmer in winter and cooler in summer. This coat of cement does have some slight insulating qualities, but it is also a means of fireproofing the roof. Most homebuilders and roofing contractors have overlooked this phase of work that is a part of the construction business. At the same time, it helps the roofing material retain its elasticity so the roof lasts longer. Old, dry roofs crack when they shake from a great flash of lightning, an explosion nearby, heavy excavation, and, especially, during earthquakes and hurricanes in districts prone to those catastrophes.

The easiest way to achieve a slurry roof is to have an established company spray your roof for added fire protection and insulation value. Some companies have their own hoppers for mixing and dispensing their version of a mixture to be used as a slurry.

You, as a homebuilder or roofing contractor, can become one of these companies. Spray your own roof as an introduction to the trade. If you have a rock roof, it's a snap to cover the surface with a slurry of cement. All leaks have to be fixed before applying your mix, because cracks are difficult to find after they are sprinkled over.

Slurry over asphalt

Asphalt shingles, because of their mineral coating, are easy to spray or sprinkle. The material sticks to the mineral like glue. It's possible to maintain the existing color, if desired. Simply add mineral coloring to the water you will be using to mix the material. Everything will have to be measured accurately and stirred well so each batch is the same—say five pounds (2.268 kilograms) of coloring per sack of cement and one sack of #30 sand.

A small amount of mix on a shingle in the hot sun will dry in minutes to give the true color of the finished product. You will soon have the matching shade by adding more color or diluting as needed.

Coloring stucco

Some exterior stucco comes already mixed dry in different colors. It's readily available at material houses in the west. Add water to the material that comes with the color in the sack. If the color comes in a box, mix this with measured water. For small jobs, mix in a wheelbarrow to a smooth consistency (no lumps).

Applying slurry

It will take but a short time to cover a section of the roof by hand with a wide brush made of cheap fiber. The slurry shown in Fig. 13-1 was applied with such a brush, splashed onto a roof covered with 90-lb. (40.82 kg) felt. You can use a regular dash brush, available from the same yard where you buy your sand and cement. Each time you dip your brush into the container of slush, push it to the bottom. This will keep the material mixed and the sand from sinking to the bottom.

Figure 13-2 shows a slurry of cement that was applied using a large water brush. Even if you use a gun to apply the slurry, you might need a brush to apply the material in a difficult area where the gun nozzle doesn't fit.

Ready-mixed material comes with a courser sand than #30, which makes it difficult to dash with a brush, and likewise difficult to spray with a light spray gun. It takes a heavy gun to shoot a slurry of ready-mixed exterior stucco.

143

Figure 13-1

Slurry applied over mineral-coated roofing with a wide brush.

Figure 13-2

Slurry applied over rock using a large water brush.

Mix your own slurry to sprinkle your roof to a uniform thickness, say ³⁄₁₆ inch (4.76 mm). The slurry can be sprinkled onto your roof in a short time if your arm doesn't grow tired.

If your roof is quite flat, you can place your mortar box on the house and mix there. Otherwise, you can carry the material onto the roof in buckets.

Using acoustic guns

Protect your shrubs from overspray with sheets of plastic (one mill), bought from the material dealer where you buy cement products. The material can be sprayed onto any roof in minutes with a hefty acoustic gun, but you might also end up spraying everything around the yard. In addition, when using a gun, the slurry has to be of an excellent creamy consistency for pumping. This is where the experts come into their own. They have the already prepared material in

Figure 13-3

Acoustic gun with hose.

sacks and are ready for any job, which would include protecting shrubbery and out-buildings from overspray.

It's not likely that you will spray wood shingles, but these are the most likely to catch fire in areas where dry brush is prevalent. If you do opt this route, make sure that the slurry doesn't clog up the flow of water, so rain is forced to run under wood shingles instead of over them. Each job is an individual matter, and you have to determine whether a slurry could be used on that job or whether it could cause leaks in the wood shingles. It's not recommended to spray wood shingles, but it might, in desperation, be done in a fire-hazardous area. The material will crack and pop off quite readily over wood, especially if walked on.

A small acoustic gun is not necessarily meant to spray cement, but, when #30 sand and a one-to-one mix (one sack of sand and one sack of cement) are used, it is possible. Add a couple shovels of lime and a gun additive to give the material a slippery consistency. Most guns should handle this if mixed for twenty minutes in a clean machine. The newer acoustic guns mix their own material, eliminating a separate machine. Avoid dry lumps, or the nozzle of the gun will clog up, and you will have to dismantle it to clean out the various parts. With pressure built up in the hose, the soupy material might explode in your face when you remove the nozzle.

Cement products tend to set up in the hoses and different orifices, so you have to clean out your gun regularly, or one day everything will clog up and you'll be out of business.

Presently I know of no one who sprays roofs. I have explained how to proceed as though everyone needs to learn the business. As a contractor, I have done my share of roofs and sprinkled many with a slurry, but I warn you of the hazards a beginner might encounter when spraying a roof with an acoustic gun.

It takes but one job to learn how to spray a roof, then as a homebuilder or roofing contractor, you have a new phase of the construction business to add to your services.

A gun like this with hoses can be rented or bought. In either case you will be given instructions on how to use it.

14

Concrete roofs

THE ultimate in fireproof construction is found in Europe, where standard procedure would include a concrete roof. European methods of building fireproof do not stop with tall buildings, but extend to single-family residences. Their poured concrete roofs are probably the main reason the European incidence of fires is so much lower than that in the United States and Canada.

It might scare some people to think of all that weight over their heads, but if you take time to think it out and understand how the roofs and floors are built, you'll realize it's perfectly safe. Nearly all modern buildings in the cities have concrete floors laced with reinforcing steel rising story after story—one of several reasons why the skyscrapers are fireproof, and why terrorists can't destroy them with a few hundred pounds of dynamite. In most high-rise buildings, a person could probably weather out nearly any fire simply by staying in one room and blocking off the smoke from coming under the door. The building is not going to burn down, only its contents. If one should go out into the hall, he might become asphyxiated from breathing poisoned air from burning fixtures and rugs.

All European countries build more fireproof than we do in the United States. They started hundreds of years ago because stone was safer against flying arrows than anything else, while tile roofs became safer against flaming arrows. In some areas, wood was gradually depleted as fuel, leaving the people little choice of building products. But lumber has always been plentiful in the United States and Canada, the reason the two countries have the greatest incidence of fire in the entire advanced world. (Of course, this is not comparing our ratio of fire damage to slum areas of poor countries where they build with straw or cardboard.)

Since the concrete roof in many European countries is installed as a matter of course, European builders don't know how to build a worthwhile house any other way. It's a matter of integrity. Builders are proud of their work and want it to last a hundred years, or even several hundred years.

Preparing for a concrete roof

Today concrete block and brick have, for the most part, replaced stone for walls, but block and brick make substantial foundations for the heavy roofs of concrete. A load-bearing center partition down the middle of a room poured solid will help support the concrete beams of a residence. Then, the forms will be set over reinforcing rising from the center wall (Fig. 14-1). In the case of block walls, you can tie in the steel reinforcing with the forms built for rafters (Fig. 14-2).

Figure 14-1

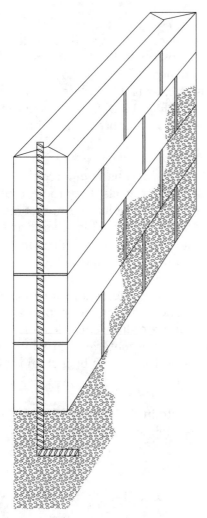

Note the rebars coming out of this concrete wall.

Figure 14-2

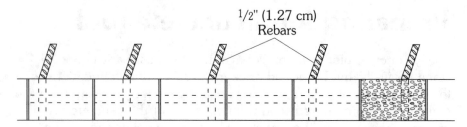

1/2" (1.27 cm)
Rebars

Steel from the block walls ties in with the forms built for rafters.

Rafters

Concrete rafters and trusses take the place of the wood equivalent prevalent in the United States. A trough is built in place for roof rafters and trusses. It might be nine inches (22.9 cm) wide and 12 inches (30.5 cm) deep, notched four inches (10.2 cm) deep on each side, and two inches (5.08 cm) wide. In Fig. 14-3, you can see the rafters, with reinforcing in place, ready to be poured solid with concrete.

Each roof designed for concrete would have to be engineered individually. If there are to be middle bearing walls, they too are of brick or block or poured concrete. A concrete column is sometimes built up from bearing walls to support the rafters. This forms a truss. Steel is placed in each trough and a lid nailed down. The trusses to be poured with concrete are shored up to support the weight.

It is best to pour rafters in place. The hole in the beam shown in Fig. 14-4 will set over a large bolt in the bond beam or roof support. Then the hole will be poured full of concrete. In France, a stationary crane is generally set up on each job and the material is lifted in large buckets. Then a gate is opened at the bottom of the bucket and the concrete is poured into the troughs with ease.

A series of rafters are formed in concrete in this manner. Floors between the multistoried buildings are built similarly, except they are flat. European countries don't use nearly as much steel to tie the buildings together as we do in the United States.

Figure 14-3

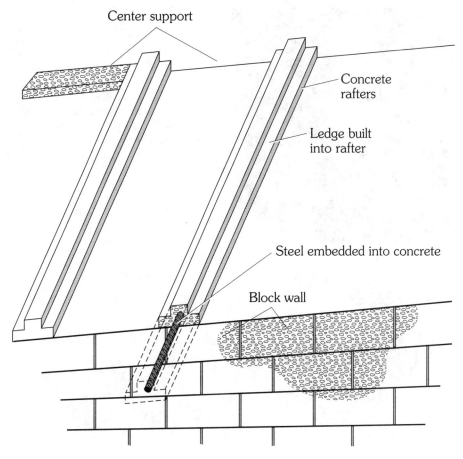

Concrete rafters, boxed in at four feet (1.4 m) OC.

Tile

When the concrete for the rafters has thoroughly set, the shoring is removed. Hollow tiles are then set in place (Fig. 14-5). The tiles are to set exactly from ledge to ledge on the rafters or floor joists. The crane then lifts a box of tiles onto the roof where they are easily set onto the ledges. A roof or floor is built in a hurry after the timely task of building the troughs for the joists and rafters.

Figure 14-4

Concrete beam with two pieces of one-inch (2.54 cm) steel for concrete rafters.

Figure 14-5

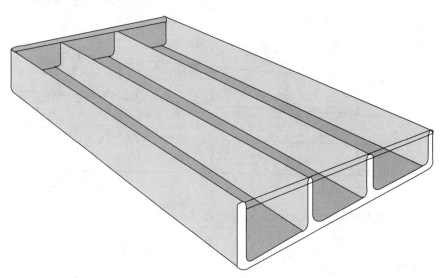

Hollow extruded tile set on the ledges of rafters.

A roof like this, covered with tile, will last several hundred years. The tile fill with silt from the air, weed and small trees sprout, and still the roof doesn't leak.

These days, a roof is covered with a layer or two of basecoat roofing material. You might even hot-mop between layers. If the roofing material does disintegrate slightly after 50 years, the concrete can easily absorb the little moisture that might hit the surface. Work has to be properly coordinated so tile is not walked on. Once the tiles are set in concrete, however, they don't readily break. All plastering, on the roof, around chimneys and parapet walls, should be completed before tiles in those areas are installed.

The roof of hollow tile is built extensively in Europe, especially in Italy, San Marino, and France. It should be used only where earthquakes are not a problem. With a shake, the tile could become dislodged from the two- or three-inch (5.08 or 7.62 cm) ledge and tumble down.

As a result of our building codes, this type of construction is not found in the United States, with only rare exceptions. The hollow tile would have to be built as in Fig. 14-5, with steel reinforcing and tied into the steel in the rafters. Here you can see that the top half has been removed, leaving the troughs to be poured solid. You can then secure ⅜-inch or ½-inch (9.5 mm or 1.27 cm) reinforcing bars to the steel coming out of the beams or rafters and tie to the steel placed in the troughs of the tile, overlapped and tied together. Then, as you cover the floor with an inch (2.54 cm) of concrete, at the same time you fill the tile. In Europe, the extruded hollow tile is a standard product, used in everyday construction of floor and roof assemblies without reinforcing steel (Fig. 14-6).

A case study

To prove that what I have set forth in this book was feasible, I designed and built two small factories with concrete roofs. The span was 22 feet (6.7 m). I set the forms for concrete beams and supported them with plenty of shoring.

Since I was building in the States under stricter building codes (for earthquake reinforcing), I was required to use adequate steel, much more than I have ever seen used in Europe. I used 2 #8s, placed in the forms for the span, three inches (7.62 cm) from the bottom and supported in the middle by chairs (reinforcing supports) where the

Figure 14-6

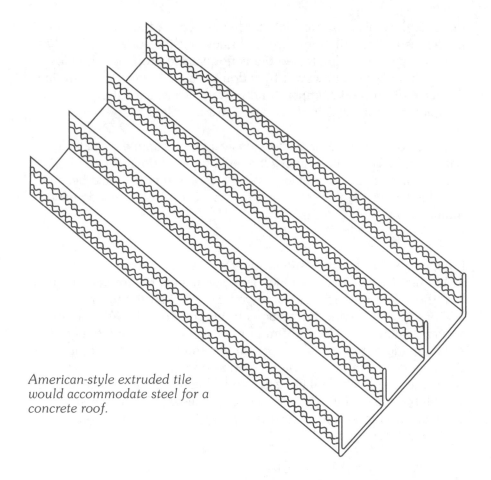

American-style extruded tile would accommodate steel for a concrete roof.

rods sagged down too far. Within two inches (5.08 cm) from the top of the form, I placed #3s horizontally, allowing them to come out of the forms about 18 inches (45.7 cm). I built the two-inch (5.08 cm) ledges into the forms for support of the concrete between beams.

I poured these forms full of concrete mixed on the job and lifted in buckets with a forklift. After a few days, I removed the sides of the form and left the underside stationary and shored up.

Since no extruded tile are available to bridge spans between beams, I used 3.4 (1.62 kg) rib lath, the most rigid of that type of material. Over the roof and over the newly poured beams, I placed #3s every

18 inches (45.7 cm) the entire length of the building, which was 50 feet (15.2 m). I laid #3s every 18 inches (45.7 cm) apart in the other direction, forming a grid of reinforcing for the roof.

Over the metal lath, I placed a few pieces of cement blocks where the rods sagged. I tied the metal lath to the rebars so the metal wouldn't sag in the middle when plastered. I fastened down the ends of the metal lath so it wouldn't wiggle when plastered from underneath with two coats of cement. My roof had a slope of 2:20, but in Europe they build gable roofs similarly so snow will slide off easily.

I was afraid the metal lath wouldn't support the concrete to be pumped into place, even though the mixture was to be lightweight. So I placed a shoring of 2×10s (5.08×25.4 cm) under each four-foot (1.22 m) panel to be poured. I pumped three inches (7.62 cm) of concrete onto the roof, floated it off, and troweled on a smooth coat of cement. I didn't expect my roof to leak, but it doesn't rain often in California, so it didn't have a test until winter. During the waiting time, a small earthquake shook the building. A few cracks developed in the concrete roof, and it leaked. Figure 14-7 shows the end product.

Figure 14-7

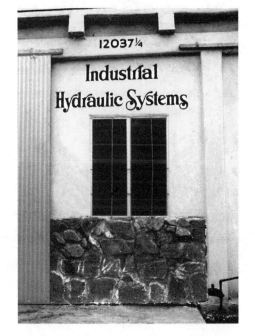

My experimental all-concrete U.S. building.

I put two layers of #15 felt over the entire roof, lapping halfway and rolling the material with lap cement. Since I tend to think fireproof, I couldn't leave the tar paper exposed, even though, on the off chance it caught fire, the building would never burn. I placed stucco netting over the entire roof with gravel guards around the edges to accommodate the stucco in a neat fashion. I then covered the roof with two layers of cement, troweling the second coat reasonably smooth. I don't ever expect to have to repair the roof.

This method of a troweled-on finish is done where the weather is nearly frost free. The practice has to be dealt with through the building department in each particular region to ascertain what frost might do to cement when used as a roof covering. Will the finish flake off? Will water accumulate under the stucco and lift it in winter? And so on.

Since the netting was installed over concrete it wasn't nailed down. The weight of the material held it in place.

This type of building is not just for the rich who spare no cost in building; it is, rather, especially for those who don't have the money to rebuild. If you live in a forested area, where fires are a threat every dry season, you would never have to endanger your life by watering down the roof while the police were ordering everyone to evacuate the area.

After the roof had been poured, and the concrete beams had to set for 21 days, I removed the shoring and shot the entire ceiling with acoustic.

Shoring

There is more to building a concrete roof than any other type of roof. The shoring to support the weight of the wet concrete is especially important as well as costly. There are companies that do nothing but this type of work; they can use their shoring several times over, passing the savings onto you. It might be wise to seek their expertise.

Heavy plywood in 4-×-8 sheets is often used for shoring, supported by 2×4 (5.08×10.16 cm) or 2×6 (5.08×15.24 cm) every two feet (0.61 m) in one direction, then when necessary, supported by additional shoring in the opposite direction. The assembly is all tied together by cross braces.

Workers have to work on this temporary floor to install reinforcing in two directions, tied together in the form of a grid. This grid is held off the face of the false floor by means of metal chairs, mentioned before, or small pieces of cement block or rock. Make sure that the grid of steel doesn't slip or you might have to lift the reinforcing back into place while pouring.

Again, this type of work has to be done by experts who pour concrete and build shoring regularly. The plans have to be engineered as well as the shoring in many instances. Plans will have to be approved by the building department and a permit to build is needed. There is, I admit, more of a hassle building this type of building. The advantage, of course, is that you only have to build once; it will never go up in smoke.

Reinforcing rods for the roof have to be tied into those coming out of the walls and should be lapped 16 inches (40.64 cm) to two feet (0.61 m), depending on the engineered plans. You might leave the cells of the block empty for about a depth of eight inches (20.32 cm) so the entire roof structure and the top of the block wall are tied together as one unit, (Fig. 14-2).

When the forms are ready, the concrete is pumped onto the roof or floor to a depth of four inches (10.16 cm) or as predetermined by the engineer and approved by the building department. If weight becomes a problem, you can turn to lightweight aggregate instead of sand and gravel. The more weight you add to a building, the more costly the walls and foundation become since they have to be designed to support the weight.

The engineer should state on the plans when the shoring is to be removed after a pour. Building with concrete, especially when the wet material will be handled overhead, is not play. Even the experts building freeways have had their forms topple.

Steel reinforcement

Fireproof construction in the U.S. can't take place without steel reinforcement, as I mentioned earlier. In Fig. 14-8, you can see what might be the ridge for a new house. To eliminate the wall in the center of a building, an I-beam can sit on columns or another poured-concrete support member. Set one end of the beam on a reinforced column with bolts provided to match a wooden template that fits the bolts in the beam. Lift the beam in place over the bolts.

Figure 14-8

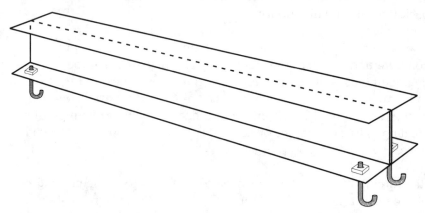

I-beam as ridge.

Next, tighten the nuts so the beam remains solid at one end. Fill the cells with concrete at the other end, and tamp round the bolts inserted in the beam. A week later, tighten the nuts on the anchor bolts set in the concrete. The steel will have to be lathed and plastered to classify as fireproof.

Concrete's durability

Concrete, known for its durability, is used in the building of bridges, dams, canals, silos, and farm buildings. For years the Santa Fe Railroad has built concrete houses for employees, and they still can be seen along the tracks, in as good a condition structurally today as they were nearly a hundred years ago when some were built.

Concrete will support more than 10,000 pounds per square inch (4530 kg) and will become harder with age. Concrete columns built 3,600 years ago are still in existence today. And the cement manufactured today is far superior to that produced in the historical past.

I helped build a tank similar to the one in Fig. 14-9 the first week after my graduation from high school. A framework of metal lath set up lengthwise and tied together, the tank was 18 feet (5.5 m) across and eight feet (2.44 m) high. No upright steel was used, but it was wrapped with #9 wire every four inches (10.16 cm). That was in the early 1930s, and the tank is still in use as a supply tank on a cattle ranch. If built today, the tank would have at least ⅜-inch (9.52 mm) reinforcing bars incorporated into the lathing.

Figure 14-9

Even though this concrete water tank with a poured-concrete roof need not be fireproof, it will last for hundreds of years.

Concrete slabs and beams are used in the assembly of parking structures. The same slab can be swung over a concrete structure for a concrete roof. In fact, many are built this way in the Middle East. This might mean a flat or almost-flat roof, but it can be made into a steep roof with a steel support beam or ridge down the middle and lightweight slabs swung into place and tied together at the ridge. A hole can also be provided in the slab or beam at the bottom and a one-inch (2.54 cm) bolt dropped in the block wall. The bolt should come almost to the surface of the hole in the slab or beam. The hole is then tamped full of half-dry cement and a little sand.

No extruded tile like the ones shown in Figs. 14-10 and 14-11 are manufactured in the United States. If there was a market for the material, tile companies would gladly manufacture them. The market will come only when we become serious about fireproof construction.

Figure 14-10

Extruded tile used in concrete roofing.

Figure 14-11

Extruded tile used in concrete roofing.

Sicilian tradition

Sicily is at its best in the construction business using cement products. The whole of the island is mountainous, but there are no trees for lumber.

Since this is a chapter on concrete roofs, I need only to say that the floors and roofs are built the same. The concrete beams for the roof only slope to shed water (Fig. 14-12).

In Sicily, there is only a narrow strip of land near the ocean, then the building program moves gradually up the mountains. The lots are small, and the houses are close together, but none will ever burn from a neighboring fire. Only the contents of a building ever burns and that is generally contained within the room.

Figure 14-12

Concrete roof in Sicily.

15

Drop ceilings

IF steel studs are beginning to take their rightful place in homebuilding, so should drop ceilings. Fewer and fewer plastered ceilings are found in commercial buildings nowadays, probably because drop ceilings with removable panels permit ready access to wiring, the sprinkler system, and whatever might be hidden out of sight just above your head. Even in residential, more people are requesting a ceiling of fiber blocks or plastic panels where neon lights shine through.

Construction

On an almost-flat roof, you have only the sloping rafters, but the homeowners might not want the kitchen ceiling to slope. So, for block or brick or any solid wall, shoot an L-bar along the two walls, perpendicular to the joists and level at the height of the proposed ceiling. Then drop 16- or 18-gauge tie wire from punched holes in the ceiling joists. Punch holes in T-bars for the tie wire and hang the bars every two feet (0.609 m) or the width or length of the paneling you will be using, parallel to the L-bars attached to the walls (Fig. 15-1).

Figure **15-1**

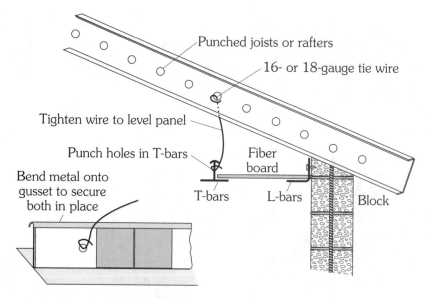

Gusset fastens T-bars together.

The T-bars come in 10-foot (3.45 m) lengths. Use flat metal gussets to fasten the 10-foot (3.45 m) lengths together, crimping the overhang on the T-bars to the gusset (Fig. 15-2).

Figure 15-2

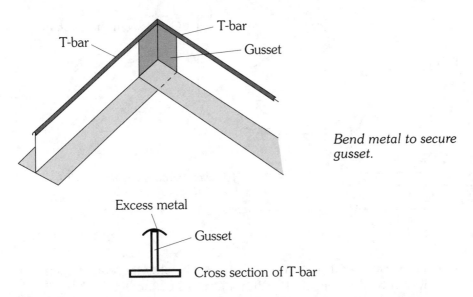

Bend metal to secure gusset.

Your paneling may come in 2-×-4 (5.08×10.16 cm) sheets. Cut some T-bars in two-foot (0.609 m) sections. Insert these every four feet (1.22 m), perpendicular to the original run to accommodate the paneling. It is best to have your paneling ready and set in a section to make sure that everything fits.

Bend a gusset into an L shape (Fig. 15-3). Insert one in each corner of the box you are forming to accommodate the paneling. With a pliers, crimp the surplus metal at the top of the T-bars to secure the gussets (Fig. 15-4).

Cutting panels

On one side of the room, the last panel will come out an odd width. It will depend entirely on the type of material that you have on how you will cut it to size. If the paneling is fiber material, use a fine-tooth blade in an electric saw (a bench saw would be best). Cut the

Figure 15-3

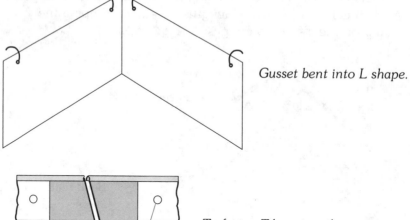

Gusset bent into L shape.

Figure 15-4

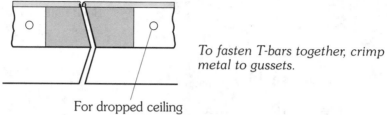

To fasten T-bars together, crimp metal to gussets.

For dropped ceiling

material to size, and place it in position. The slightly ragged edge will be covered by the ¾-inch (19.05 mm) return of the T-bars or wall angle.

If you have light panels for the neon light to shine rough, you will use obscure plastic sheets under those sections. When a full section of paneling does not fit, you might have to score the material with a sharp tool and break it. If the edge is slightly oversize, break off the little jagged pieces with a pliers or wire cutter until the panel fits.

You can install the T-bars for any size paneling and you will have a floating ceiling. Fiber board comes in many sizes and textures and gives better insulating value than does obscure plastic sheets.

This type of ceiling assembly would be fireproof, except for the paneling. Just remember that the more elements of a house that are built fireproof, the less there will be to burn in case of a fire.

Helpful hints

If your ceiling is a high gable, you need only use longer hanger wires for the same effect, and heavier wire when needed according to the inspector or the building codes. If you will be doing any great amount of installing this type of paneling, be sure to buy a stone of soft wire that will loop and bend more readily than the stiff tie wire used as a backing for paper in stucco work.

The one drawback to this type of paneling is that it has a tendency to retain odors. Unpleasant odors can permeate the attic above the kitchen, for example, and invade the adjoining rooms. As an alternative, the ceiling of the room most likely to retain unpleasant odors can be plastered. This would entail the use of heavier hanger wires and metal lath screwed to the T-bars (all, of course, with the permission of the building department) so the ceiling is sturdy enough to remain intact. In the attic, the framing between rooms can be plastered to form a firewall above the room in question. The solid covering will prevent the odor from spreading. Wallboard might suffice, providing all the joints are thoroughly covered with one or more applications of joint compound on both sides of the wall.

If there is already a firewall, it might only need to be built odor-proof. If odor can penetrate a firewall, so can a fire. Just work the wall over with joint compound until no air penetrates. A small acoustic gun, if the nozzle reaches the area causing the trouble, would only take five minutes to blast the wall with some type of material that would shut out all odors.

16

Tile roofs

IN most European countries, the roof of a home is expected to last the lifetime of its owners. Many concrete roofs covered with tile might even last the lifetime of the building. Many 200-year-old houses in France are being renovated today and are receiving their third tile roof, while we in the United States are continuously re-roofing old buildings several times within 50 years. Asphalt applications, when exposed to the elements, will gradually deteriorate. On the other hand, today's tile, when heavily glazed, should last through the ages, especially when applied over a concrete roof (as is the norm in many European countries).

When I built my fireproof house, I was forced to proceed with a conventional roof. I used ½-inch (1.27 cm) exterior-grade plywood, screwed to C-studs used as rafters every 12 inches (0.3048 m)—six inches (15.24 cm) along the edges. Over this I applied a layer of wallboard to negate the fire potential of the plywood. (I was hoping a fire would never reach the interior of the house.)

I then added two layers of 30-pound (13.61 kg) felt and nailed that progressively with BRT A/T square cap 1½, about an inch (6.45 sq. cm). This type of nail is necessary where a 90-mile-an-hour (145-km-per-hour) wind might blow up. Obviously, you need only follow the nailing requirements in your area. If you are going to cover the roof with tile, you probably can dispense with these rigid nailing requirements.

The battens in most cases will be nailed down, or at least the tile will be nailed down and the sheer weight of the material (9.2 pounds (4.12 kg) per square foot (0.3048 m)) should hold the roofing in place even in the event of a great wind.

Next, I hot-mopped the roof progressively and applied rock. Using a light acoustic gun, I shot a slurry of cement over the rock. This was intended to negate partially the potential of the combustible roofing material in all instances, except from an extremely hot fire. The slurry of cement also kept the roofing material from losing its oils and the rock from blowing off the roof and into the pool.

History of tile use in U.S.

When the padres on the West Coast built their first durable buildings, they thatched the roofs with rushes and reeds. This practice didn't last long after the first attack by a hostile tribe of natives, who shot fire arrows into the dry material.

The padres turned at once to the Mediterranean-style roofs used in their homeland. They fastened the tile over limbs placed horizontally over the support timbers. Because the tiles were fireproof, the problem of roof fires was eliminated. We, too, can learn from the Europeans (Fig. 16-1).

Figure 16-1

In San Marino, the buildings are sometimes close enough that one can jump from one to another. With all tile roofs, the fire damage is far below that in the United States.

Tiles are made similarly to bricks, from baked clay, under tremendous heat and cooked so they do not later disintegrate. This is the best fire-prevention roof known in the civilized world today. Tiles are made by pouring the *slip* (liquid clay) into molds where it is allowed to dry and is then fired. Then the more modern method is to make tiles by *extrusion*. This method forces the wet material through a mold. The tiles are cut off, all the same length. The different shapes of tiles are installed in different ways. A layer of concave tiles is placed on the roof, covered by a layer of convex tiles.

Also available are flat tiles, which are installed like shingles. The red, yellow, and varied similar colors result from the nature of the clay—

the amount of iron present—but cement tile have the color mixed into the material.

In the United States, we move a crane onto a job for a day when one is needed, but in Europe a long pole is buried in the ground with a crane attached. This remains on the job until the house is completed. This type of crane is cheap rental compared to the movable crane on wheels (Fig. 16-2).

Figure 16-2

A stationary crane is installed on the job for bucketing up concrete for the all-concrete roof and for lifting up boxes of tiles.

Tile has since been used extensively in the Spanish-influenced states of the U.S. Some of the more crude-looking types are still called *mission tile.* Other varieties have since been developed.

Cement tile has especially come into vogue. These are available in different colors. On the other hand, clay tile, in which the color has been baked at the time of manufacture, might retain their brilliance better than mineral coloring mixed with cement products. Plastic tiles are even available on the market, but I have a low opinion of anything that might burn.

Preparation for tile

Before a tile roof is installed, apply at least one layer of type-30 asphalt felt or organic felt meeting U.B.C. Standard No. 32-1. Carry this over or under hip and ridge nailer boards and lace it through the valleys.

If the roof pitch is below 4:12, you need to install an approved built-up roof over solid decking to comply with the codes. Nail 1-inch-x-2-inch (2.54×5.08 cm) wood battens, vertically over the built-up roofing at 24 inches (0.609 m) OC and nail or screw them into the rafters. Hot-mop the entire roof. Then, nail horizontal battens to the vertical battens at not more than 14 inches (35.58 cm) OC, depending on the length of the tile that you might choose. Battens should have ½-inch (12.7 mm) separation at the ends for expansion.

Tiles generally have one or two nail holes along the upper surface so they can be attached to the battens. With regular roof slopes, skip the vertical battens and apply only horizontal battens. Nail the tile to the horizontal battens (Fig. 16-3). The different lugs on the tile slip in back of the battens to keep them from sliding. With the tile held in place by the lugs, not all the tiles need to be nailed. Because the tiles are not rigidly nailed down, they will have a slight leeway that allows them to move during a great shake or wind and will give slightly rather than crack from ice and freezing.

Figure 16-3

The lugs on the back of the tile catch on the upper edge of the battens so that it is not necessary to nail every tile.

When using tiles for a roof, you have to make an allowance for the additional weight. The best time to think of weight is when the building is being designed. Where 2-×-4 (5.08×10.16 cm) 20-gauge studs would have sufficed for the ordinary roof covering, you now have turned to 2×6 or 2×8 (5.08×15.24 or 5.08×20.32 cm) 20-gauge or heavier when tiles are used. This problem has to be recognized and engineered for each separate set of plans.

The padres depended on tile to keep the rain out of their dwellings, but an occasional gust of snow or blowing rain entered along the sides. Today, a layer or two of asphalt roofing is required before the installation of tile. Felt with interwoven fiber or other hard-to-tear material is recommended.

All flashing should be at least 28-gauge galvanized steel. The valley flashing should have a one-inch-high (2.54 cm) splash diverter. Copper is recommended if it is to last the life of the tile, but not all material yards stock copper. Each succeeding flashing should overlap the next a minimum of four inches (10.16 cm). Do not nail into the flashing itself, but secure the metal with clips or nails bent over along the edges.

Tile installation

Tiles are almost everlasting, provided no one walks on the roof and breaks the material. Only after 30 or 40 years will you have to worry about some disintegration. At that point, you might have to remove the damaged tiles, re-mop the roof, and replace the tiles. Nevertheless, tiles are recommended for fire-hazardous situations.

To avoid the need for anyone walking on the roofing tiles after they are in place, you should place them only after all the exterior work or plastering on the roof has been completed (the parapet walls or around chimneys, for example). If the tiles are installed, the roofer has to at least leave a path for workers to complete the work that should have preceded the installation of tile.

The roof has probably already had two or three layers of felt applied and mopped, so there will be no immediate leakage caused by damaged tile, but the area will disintegrate over the years faster than that not exposed to the weather.

To install tile over steel joists, be sure to first box in the eaves. With a pin gun, fasten ⅞-inch (22.22 m) Milcor to brick or block walls as a plaster ground. There should be a bond beam in that location. You might be able to use cement nails to secure the Milcor. This installs the plaster in a neat fashion, instead of directly against the wall where it will show a crack from the constant shaking at the juncture of the ceiling plaster. You might use the track as a fascia, another way of breaking off the plaster at the outer edge. You will have to screw the metal lath to the ends of the joists, looping the 18-gauge tie wire through the punched holes in the Milcor to fasten the rest of the metal. Make sure all webbing is up, to facilitate plastering (Fig. 16-4). The tile should project over the eaves to prevent the discoloration of the walls from water. Even better, include a rain gutter to catch all rainwater and divert it to the ground.

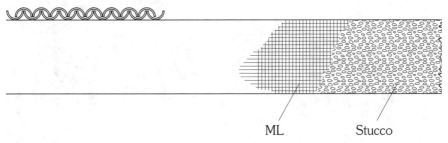

Figure 16-4

ML   Stucco

Fascia board should be plastered to prevent continuous painting of the eaves.

In Fig. 16-5, you see a roof ready for the installation of the first row of tile. Here, individual blocking was required to keep sparrows out. When long strips of blockage are used, they also serve to align the tile for straight rows. As you can see, not all tiles are nailed to battens. Incidentally, the weight of the tile placed on the roof in advance helps to settle the building. The false fireplace has to be lathed and plastered before the tile work is completed.

Most of Ireland has turned to tile, either clay or cement. Some districts use mostly slate where that is plentiful and nearby. All

Figure 16-5

Long strips of blockage also serve to align the tile for straight rows.

Figure 16-6

Example of Irish flat tiles.

buildings in Ireland are quite fireproof. Ireland has scarcely any wood products, except for what may be harvested in years to come. Imported wood shingles are expensive, so builders naturally turn to tile (Fig. 16-6).

Figure 16-7 shows a home all ready for tile and truly fireproof—all steel studs, fireproof stucco, and a tile roof.

Figure 16-7

This modern, all-steel-frame house in the U.S. will receive a tile roof.

Fireproof partitions

UP until now, we've talked about building the exterior walls of a structure and putting on a roof. Next comes the installation of interior overhead wiring (all contained in thinwall conduit, especially when farther access to the attic will not be possible). The lathing contractor is then ready to lath or apply wallboard to the ceiling. Of course, this is best done before the homebuilder has partitioned off the different rooms, which would mean much less cutting of lathing materials. But what's the best partition method to use?

Two-inch (5.08 cm) solid partitions

When remodeling, the average homebuilder will turn to 2-x-4 (5.08×10.16 cm) partitions as the cheapest and easiest to assemble. But anyone who has been in the building business for any length of time will have run into the two-inch (5.08 cm) solid partition, far easier to install and cheaper than anything in wood. With ¾-inch (19.05 mm) channels or ⅜-inch (9.52 mm) reinforcing as studs, you lath one side with metal lath, while you plaster both sides to attain a straight and true wall, approximately two inches (5.08 cm) thick, and fireproof.

This type of construction has been used extensively for years in multistoried structures because of the weight factor. There is not much to the wall, hence the savings in weight, especially if you use lightweight aggregate such as perlite or vermiculite when plastering. But the main reason for turning to solid partitions is the fact that they are entirely fireproof. The two-inch (5.08 cm) wall has seldom been incorporated into residential construction, but there is no reason why it shouldn't be used there more extensively. This method of construction is especially suitable for short runs, closet walls, etc.

If it's difficult to find a steel door frame for a two-inch (5.08 cm) partition, use a 2×2 (5.08×5.08 cm) and a finish door frame. Split and saw this in two lengthwise. Attach metal lath to the rough frame with long nails as shown in Fig. 17-1. The finish door frame will form a plaster ground. After plastering is completed, cover the rough frame with a wide trim. The lock set can be attached to the rough door frame. Place an extra dowel in the slab next to the 2-x-2 (5.08×5.08 cm) rough frame when necessary and fasten the two

Figure 17-1

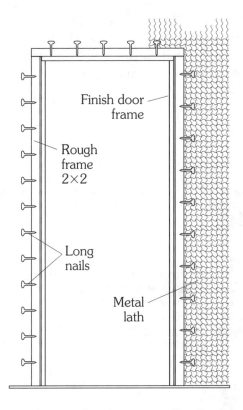

Finish door
frame

Rough
frame
2×2

*You can improvise a steel door
frame for a two-inch (5.08 cm)
partition.*

Long
nails

Metal
lath

together so the door or wall will never slide. Two-inch (5.08 cm)
door frames are now more readily found at material yards, but you
can order them, if necessary.

Education is the key factor in promoting any fireproof job,
especially the solid-plaster wall. Since I learned this method of
construction in multistoried buildings in the Los Angeles area where
combustible materials are scarcely allowed, I have incorporated the
two-inch (5.08 cm) wall into my building program and will continue
to do so in residential building.

Metal lath

High-rib metal lath is quite rigid and forms a good partition without
upright supports, channels, or rebars and is legal, according to the

U.B.C., to 12 feet, 6 inches (Fig. 17-2). This simple partition is called a *curtain wall*. You wire or screw it in place if attached to a metal ceiling joist, or nail it in place if attached to wood joists. Use 8d nails, driven three-quarters of the way in; then bend them over, securing several webbing of metal.

Figure 17-2

A sheet of rib lath attached to a steel joist to form the beginning of an inner partition.

A sheet of rib lath is two feet three inches (0.6147 m) wide and eight feet long (2.44 m). It is manufactured to be installed over studs at 16 inches (40.64 cm) or 24 inches (60.96 cm) OC, with a two-inch (5.08 cm) overlap. Allow the sheet of metal to rest on the floor; you might want to shoot a pin into the floor to hold the bottom where it belongs. If you are working over wood flooring, a 16d nail or two will suffice to hold the wall in place.

All metal lath is lapped two inches (5.08 cm) and perhaps three inches (7.62 cm) lengthwise or at the ends. The metal is tied together every six inches (15.24 cm) with 18-gauge tie wire. The ties are mashed flat for easier plastering.

This type of wall may seem flimsy, but it can be plastered on the smooth side first and will be quite rigid for the second coat. Broom the first coat to facilitate adherence of the second application of plaster or cement. After the material sets, apply a second coat. The wall is plastered in successive coats of cement or plaster to a minimum thickness of two inches (5.08 cm), including the finish coat. Straighten the walls with a darby or rod. If necessary, use a level to plumb the walls for an application of tile.

You might find that some walls come out a fraction more than the required two inches (5.08 cm) to form a straight wall (Fig. 17-2).

In Fig. 17-3, you can see dowels embedded in concrete, either during the process of pouring new concrete slabs or by drilling holes in the old slab at 16 inches OC (40.64 cm). Notice how the rebars are secured to these pegs to form the stud wall. If you are using channel iron, drive a 10-inch (25.4 cm) section into the drilled, half-inch (1.27 cm) holes in the concrete. If you are working over a wood floor, saw a section into the channel at four inches (10.16 cm), and bend it into an L shape. Drill the necessary holes into the foot, and nail this to the floor as an

Figure 17-3

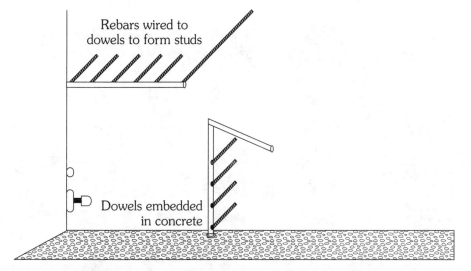

Dowels embedded in concrete, either during the process of pouring new concrete slabs, or by drilling holes in the old slab at 16 inches OC (40.64 cm). Note that the rebars are secured to these pegs to form the stud wall.

183

attachment for the studs to be installed or to form the bend in the stud and install. Wire metal lath to the newly formed wall using #18 tie wire.

If you are lathing an extensive ceiling with metal lath, mark the floor where partitions are to be located, and drill holes for the pegs that are to start the walls. Finish lathing the entire ceiling before installing the walls. With a level and or plumb bob, snap a straight line across the newly installed ceiling of metal lath. Mark off the different rooms and halls. Poke the rebars as studs through the metal lath along the line on the ceiling. Fasten the rebars to the proper peg in the floor. In minutes, you have a wall ready for lathing. The rebars do not have to be exactly perpendicular except when next to an opening.

Wire the studs to the pegs at the floor line using 16-gauge tie wire, doubled and in two places. If you have only 18-gauge tie wire, attach with another loop or two.

Be sure to lap all metal in the corners, and in the ceiling angles if they are to be plastered, so you will have a continuous reinforced room. You can buy metal angles for this purpose and wire them into place (Fig. 17-4).

Figure 17-4

Drill for fastening metal to steel joists with approved screws.

Drill for fastening metal to steel joists with approved screws. The material house will provide the correct gauge, perhaps wafer-head Streaker #2 Phillips recess. It is sometimes a good idea to check with the building inspector to make sure he agrees. If one type of screw is not available, another will be allowed.

Fasten the screws through the heavy sections of the metal lath, not through the lighter webbing. Make sure the corrugations are upward, the smooth side down for easier plastering. For corners or angles, use a Skilsaw with a metal blade for cutting rebars and channels. When no other method is available, wire in place with 18-gauge tie wire. Use a hacksaw to cut a slot for bending the foot of the channels.

If rib lath is applied straight, it will fit together in a groove similar to tongue-and-groove flooring. On the ceiling, attach every six inches (15.24 cm), but if it is fastened to the heavy part of the material as recommended, you will need to attach it with screws every five inches (12.7 cm).

Using a wide head screw, it should cover sufficient webbing to keep the ceiling intact in an emergency.

At one time, metal-lath ceilings were wired in place. With screwable studs and joists and the screw gun, this phase of work has become much simpler.

When metal-lath ceilings become detached in a disaster, the plastering material remains attached, and the sheer weight of plastering materials can yank even more of the ceiling down. In this way, an entire ceiling can be pulled down by the weight of the plastering material. For this reason, be sure to use a wide-head screw for proper attachment of metal lath for ceilings of large areas. A long time ago, washers were applied over each nail when metal lath was fastened to the wood joists of schoolrooms. These washers became known as *schoolhouse washers*.

I was once given special permission by the local building department to use ⅜-inch (9.52 mm) rebars in place of ¾-inch (19.05 mm) channels. I've since learned that ⅜-inch (9.52 mm) rebars are part of the U.B.C.

Poke the rebars or channels, cut to length, through the metal ceiling, penetrating 2 inches (5.08 cm). You have your room framed for fireproof walls. If you don't think the walls sufficiently secure, you can dump a little plastering material on each reinforcing rod or channel running into the attic.

Tie your metal to the rebars every six inches (15.24 cm) or to the channels serving as studs. Run all metal horizontally, lapping at least two or three inches (5.08 or 7.562 cm) along the sides and ends.

Cement plaster

When plastering inner walls, you can use cement plaster instead of interior plastering material, such as Hard Wall. Interior plastering material will not hold up under damp conditions for any length of time. Plastering with cement costs less since you can add more sand than you can with Hard Wall and not damage the quality of work. Then, if you should have a leaky pipe, cement will not deteriorate in dampness. Cement can be troweled smooth the same as any finish plastering material. A little lime is allowed; added to the mix, it facilitates workability.

All finish coats to be troweled smooth should be half sand (#30) and half cement for easy spreading and troweling. Keep the undersurface damp, and trowel the finish coat several times or until the surface hardens. This will prevent most checks caused from the rich mixture necessary to make the material workable. Cement walls have a tendency to crack more readily than walls where Hard Wall is used. Depending on your situation, you might turn to interior materials instead of cement products, except where water is prevalent. The building codes define the situation for your area; you will be expected to comply with their regulations.

It might not be important to use cement in residential construction, but it is essential in commercial and industrial. You may lease the building for a photo lab, where they slop chemicals around until almost any wall is eaten up. The walls will soon deteriorate. If you walk into an old store converted into a laundromat, you might see walls crumbling; in these cases, cement should have been used. Nothing but cement will withstand the rigors of continuous moisture. Newer laundromats are designed to place the washers in the middle of the store, away from the walls, preventing this water problem. But cement walls are a good idea, since there are generally spills and overflows to contend with in many types of commercial businesses. Certainly, in residential situations, cement will also stand up better when confronted with a fire.

When the lathing has been completed and inspected, the plastering contractor will then plaster the smooth side of the metal lath first. After the plaster or cement has set and the proper waiting time has been allowed between coats (according to the building code in your area), plaster the back side; if possible, fill out to the depth of the reinforcing bars or channels. The wall will probably require one more coat on each side to bring it true and straight before the finish coat is applied.

When completed, the walls should be a minimum of two inches (5.08 cm) and will put no weight on the ceiling joists to which they might be tied. Even though Fig. 17-5 is not a bearing wall, the partitions will help support the ceiling joists. No foundation was required under the inner partitions. But I knew the approximate weight of the walls from previous experience to be approximately 20 lbs. (9.072 kg) per square foot (0.0928 sq. meter).

Figure 17-5

The back side of a solid partition. The smooth side has been plastered one coat.

You might think this a rather flimsy way to partition a house, but the walls stand up nicely, either with reinforcing or without. Walls with reinforcing rods are less flimsily assembled than curtain walls, making it easier to plaster. When applying the first coat of plaster, don't press too hard with the trowel or the material will fall through onto the floor on the back side. Use the trowel almost on edge, and push upward with little pressure.

Any time a phase of a construction job grows beyond your capabilities, hire an expert—a lather, then a plasterer. Or turn the job over to a knowledgeable subcontractor to handle the work.

All the interior walls in the house where I live are built with two-inch (5.08 cm) partitions, and there has never been a crack after 11 years. If this chapter is somewhat in detail, it is because few homebuilders, lathing contractors, or plastering contractors are familiar with the work done in multistoried buildings. Even most building inspectors have never seen two-inch-solid plastered walls, except within the inner cities. At one time, one building inspector said he was going to have to read up on two-inch partitions to see whether or not I was cheating him.

Drawing up plans

To keep my finger on the pulse of the building industry, I recently bought a lot. I contracted with an architect to draw up the plans. I contacted two companies that furnish trusses and studs for the job. After a month I had not had a quote on how much the engineering of the plans would cost, and only one materials company had responded. I then decided to have the engineer who engineered my house, resolve the problem. I was sure he would do the work in little time and send me a bill. I began to think I should have had him do all the work at one time, when he did the required engineering for the two-inch partitions.

I mentioned solid partitions to the architect drawing up the plans. I wrote him a sheet telling him how to approach the subject. Rather than waste our time, he said we should talk to the city plan checker.

He had never heard of a two-inch partition. He said he wouldn't give two cents for such a partition because he couldn't drive a nail into the wall. I thought fireproof partitions should be more important than driving a nail into the wall.

(Incidentally, if you want to hang a picture, drill a small hole in the plaster, slanting the bit down slightly. Drill just beyond the metal lath. In this way, any weight on the nail driven into the hole will push up on the metal lath which is very durable and stationary. You can then hang your pictures on the wall at will.)

So far, I had contacted the architect, then the engineer to furnish sheets on the ⅜-inch (9.52 mm) studs, then a separate professional for the environmental report, then someone to engineer the trusses, and finally the building department. The building department, after a preliminary plan check that cost nearly $400, asked for the walls to be engineered, for wind sheer, etc.

Since I decided to build a house in my town to facilitate writing this chapter on fireproof construction, I was forced to have this type of wall engineered (Fig. 17-6) to clarify how the wall would be anchored at both the top and bottom. The sheet was engineered by Robert Coulson, who had, some 11 years before, engineered the steel studs in the house where I live. No engineering was at that time required on the two-inch (5.08 cm) partitions.

To resolve the controversy of two-inch fireproof walls, I acquired the Gypsum Association's 14th edition of two-inch walls and interior partitions of noncombustible material (April 1994, ICBM). The two-inch-solid plastered partition is an excellent way of building fireproof and needs to be recognized by building departments and more homebuilders. The plasterer's union would like to promote the method and furnished me with more material than will be used here.

To resolve the problem of proving two-inch partitions earthquake proof, my engineer drew up a sheet, which is provided here (Fig. 17-7). One uses 3½-x-25-gauge (8.59 cm×25 ga.) steel stud blocking between the trusses and fastens them in place. A hole is provided in the blocking and a reinforcing bar shoved through and fastened at the top with a loop of wire.

Figure 17-6

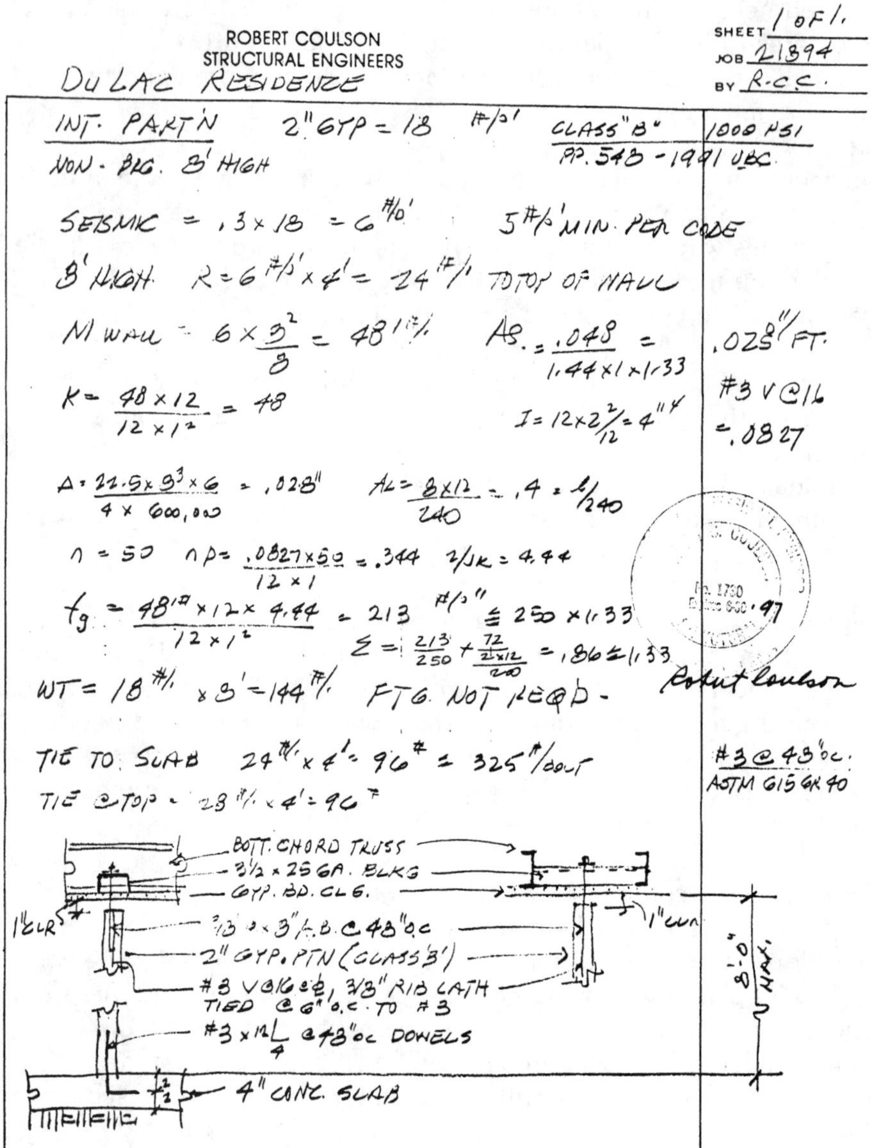

The handwritten engineering sheet reads:

ROBERT COULSON
STRUCTURAL ENGINEERS
DuLAC RESIDENCE

INT. PART'N   2" GYP = 18 #/₀'   CLASS "B"   1000 PSI
NON- BRG. 8' HIGH   PP. 543 -1991 UBC.

SEISMIC = .3 × 18 = 6 #/₀'   5#/' MIN. PER CODE

8' HIGH   R = 6 #/' × 4' = 24 #/' TO TOP OF WALL

M WALL = 6 × 3²/8 = 48'#/'   AS = .048/(1.44×1×1.33) = .028"/FT.

K = 48 × 12/(12 × 1²) = 48   I = 12 × 2²/12 = 4"⁴   #3 V @ 16 = .0827

A = 22.5 × 9³ × 6/(4 × 600,000) = .028"   Ac = 8×12/240 = .4 = L/240

n = 50   np = .0827×50/(12×1) = .344   2/jk = 4.44

fₛ = 48'# × 12 × 4.44/(12 × 1²) = 213 #/□" ≦ 250 × 1.33

∑ = 213/250 + 72/(2½×12/20) = 1.86 ≦ 1.33

WT = 18 #/' × 8' = 144 #/'   FTG. NOT REQ'D -

TIE TO SLAB   24 #/' × 4' = 96# ≒ 325#/DOWEL
TIE @ TOP - 28 #/' × 4' = 96#

#3 @ 48"O.C.
ASTM G15 GR 40

Robert Coulson

BOTT. CHORD TRUSS
3½ × 25 GA. BLKG
GYP. BD. CLG.
⅛ × 3" A.B.C 48"O.C
2" GYP. PTN (CLASS 'B')
#3 V @ 16 O.C, ⅜" RIB LATH
TIED @ 6"O.C. TO #3
#3 × 12 L/4 @ 48"O.C DOWELS
4" CONC. SLAB
1½ CLR
1" CLR
8'-0" MAX.

Sample engineered plans for two-inch (5.08 cm) partitions.

Figure 17-7

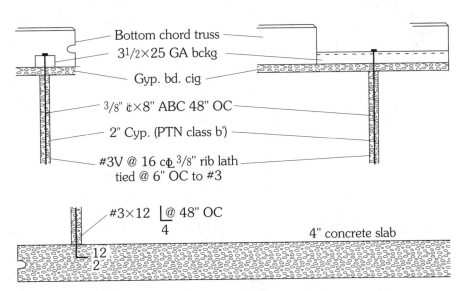

Bottom chord truss
3¹/₂×25 GA bckg
Gyp. bd. cig
³/₈" ¢×8" ABC 48" OC
2" Cyp. (PTN class b')
#3V @ 16 c¢ ³/₈" rib lath
tied @ 6" OC to #3

#3×12 └@ 48" OC
4

4" concrete slab

└12
2

*Two-inch plastered partitions using ⅜" (9.52 mm) rebars as studs,
engineered.*

18

Stucco

THE most extensively used wall covering in the western part of the United States is stucco. It was with a sneer that Frank Lloyd Wright referred to stucco. Since he generally designed fancy buildings or dwellings, he didn't use stucco extensively. Many times, he recommended a wood siding—something he considered a superior wall covering, but one that in various climates needed a coat of paint every year. Brick, block, and poured concrete are the superior products, but stucco is a utility-grade material, fireproof in itself, with a one-hour rating. It is also relatively inexpensive, so it is frequently used when other fireproof materials might be cost-prohibitive.

Stucco should not be used in wooded or brush-covered areas (anywhere with a high fire potential), but it is an excellent choice where houses are not too close together and the threat of fire is less prevalent.

In Europe, stucco has been used for many years. In Sweden, they have used their version of applying stucco for perhaps 200 years or more. They built with lumber—trees surfaced four sides and laid flat, one on top of the other, a superior log house. The trees were eight or ten inches (20.32 cm or 25.4 cm) thick, perhaps a foot (0.3048 m) in diameter. When all exterior walls were completed, they bored holes in the lumber every few inches and drove hard wooden pegs into the logs. These projected out approximately one and a half inches (3.81 cm). Over this, they applied two inches (5.08 cm) of cement in successive coats. This version of stucco was quite satisfactory at the time. The hard cement surface helped to preserve the wood and kept most of the rain and snow from damaging the vulnerable wood. It prevented insects and rodents from attacking the lumber, as well as filling in all voids between logs to keep out drafts. The hard, lasting surface slowed the cold from reaching the interior of the house. No finish coat with color was applied as we think of stucco today. The application of stucco in Sweden has advanced to the cement-block building, with a coat of cement applied the same day the wall is completed, before the scaffolding is removed.

Stucco, an application of cement, has been applied over brick buildings throughout the world, ever since it was discovered to be superior to mud, and this practice is still in use today. Cement will cope with the weather conditions and hold the building together. The application of a finish coat with color was not heard of until recently in Europe.

In France, they have learned to apply cement with a gun, and many old sandstone buildings are now being brought up to date (Fig. 18-1).

Figure 18-1

This 200-year-old house in France was being remodeled when I was there in 1993. It will receive a coat of gun-applied cement plaster. The original stone trim around all doors and windows may or may not be retained, but covered over. This building was, at one time, dashed with a coat of exterior wall covering, by hand, using a brush. The same application is now accomplished with a machine. Modern windows will very likely be installed.

The gun fills in large holes where a stone is broken away or partially missing. Then, if the owner has sufficient chips, he can apply a heavy finish coat in some form of cream or yellow, another gun application. This coat is sometimes textured with a slight troweling or left with pimples, peculiar to all gun applications. This heavy application of a finish coat of stucco with a gun seems to withstand frost better than the thin application applied in the United States.

The secret of keeping stucco from pinching off the undersurface as a result of frozen moisture is to keep the walls dry, so no moisture can form between the finish coat and the brown coat to freeze, swell up, and push off the finish coat. For this reason, finish coat should only be applied when the weather is mild and there is little chance of frost. In Europe, they have arrived at a heavy finish coat that holds up better in freezing temperatures, so the application of stucco can move farther north. The use of stucco is found even in Canada and Alaska. Because the Europeans use little lumber, however, they do not have to face the task of applying stucco to wood or metal framing, as we in the U.S. do.

Lathing

California stucco, as applied in the west, is an application of cement over wire netting nailed on over felt, or *plaster craft*, a lighter more practical version of tar-saturated felt, but not necessarily better. The netting to be applied is 1½-inch (3.81 cm) webbing, 18-gauge mesh. The tar paper as a backing keeps moisture from entering the building (Fig. 18-2).

Since the application of stucco, California-style, is not universally practiced, the method needs to be outlined in detail, beginning with the application of paper and wire as a backing for the stucco. Paper-backed wire comes in sheets approximately 3×8 feet and is easier for a beginner to apply. Look for directions on the package. You will almost need an expert on the first job to make sure the paper sheds water, or moisture might enter the house even after plastering. Cement alone does not keep all moisture out of the structure unless it is several inches thick (Fig. 18-3).

Figure 18-2

Walls covered with 1½-inch (3.75 cm) stucco netting over waterproof paper. Sisalcraft under the windows, but over the paper to shed water.

First drive 1⅛-inch (28.57 mm) board nails with ⅜-inch (9.52 mm) heads every six inches (15.24 cm), halfway into the lumber vertically at the ends of an exterior wall. To this, attach 18-gauge tie wire to the first nail individually. Walk with your tie wire, and wrap the wire around the first nail at the other end of the wall. Drop down to the next nail, wrap the wire around that nail, and walk to the other end of the wall. Wrap the wire around the next nail, until you have wire every six inches (15.24 cm) covering the entire wall.

Figure 18-3

Paper-backed wire.

To secure and tighten the tie wire, drive the nails at the ends into the two walls. Move two studdings from either end, raise the wire upward with a nail in the left hand, and drive it halfway in with the right hand. Do this with each wire at that stud from top to bottom. Skip two 2×4s (5.08×10.16 cm) and push down on the tire wire with a nail in hand to tighten it, driving the nail halfway in. Continue this until all wires on the wall are tight enough to sing, alternating up, then down on the various studs. After the wire is tight, hammer all the nails down to secure the wire.

Over this, apply 15-pound (6.80 kg) felt or some type of plaster craft made especially for lathing. Then you apply the stucco netting. The ceiling will receive 3.4 metal lath (1.54 kg), which is heavy enough to span two feet (60.96 cm). With framing at 16 inches (40.64 cm) OC, you can use a lighter lath. Lath along the walls and ceiling will lap for

Figure 18-4

The ceiling will receive 3.4 (1.54 kg) metal lath.

continuous reinforcing, or a special corner rite (bent strips of metal lath) can be used for the lap (Fig. 18-4).

Nowadays a weep screed is provided at or below the foundation plate line on all exterior walls (Fig. 18-5). The screed should be a minimum of four inches (10.16 cm) above grade and of a type that will allow water to drain to the exterior of the building. The weather-resistant barrier and exterior lath should cover and terminate on the

Figure 18-5

Weep screed, applied so moisture drains away from the felt backing at the foundation or slab level. With this separation, moisture never creeps back onto the stucco.

attachment flange of the screed. (See Universal Building Code, Sec. 4706 (e) Application of metal plaster base.)

When the windows are set in place, place Sisalcraft behind the windows. It comes in rolls made especially for around window and door openings, to keep out water. Starting at the bottom, place a layer of Sisalcraft so it is held firm when the window is shoved in place. Do the same along the sides and top of the window. In Fig.

Figure 18-6

Sisalcraft in place.

18-6, you can see the Sisalcraft hanging down, crumpled by the wind. It will be straightened out, covered with felt, and fastened down with stucco netting. This is more difficult to do with steel studs and a screw gun than with a hammer, nails, and wood framing.

After the walls are covered with tie wire, roll out 15-pound (6.80 kg) felt or a plaster craft paper in sheets that you can handle. A good lather will handle the whole roll of plaster craft and nail it at the top as he rolls out the material. It used to be that one lapped three inches (7.62 cm) over the foundation. Now, you simply fit the plaster craft into the groove of the weep screed, as shown in Fig. 18-7.

Figure 18-7

Weep screed applied.

Plaster craft is made especially for lathing preparatory to stucco. As you roll the paper out, hold it tight. Drive a nail occasionally along the top to secure the craft. Roll out a second row of paper, lapping the second three inches (7.62 cm) over the first, so it sheds water like shingles.

Over this, roll out your stucco netting and nail it in a few places. Stucco does not keep out the rain, the paper does, so a good lath job is mandatory.

After two rows of wire are applied, nail it down every six inches (15.24 cm), at every 2×4 (5.08×10.16 cm) or 16 inches (40.64 cm) OC. Stucco netting is now made with a crinkle so the netting is self-furring, which means the cement will be under the wire as well as over the wire.

Use a 1½-inch (3.89 cm) galvanized nail. Nowadays a staple gun is used if you have any appreciable amount of work to be done. The gun is operated by air from a compressor. Many different types of nailing guns are available today, and most are quite good. A lathing contractor might want to have more than one so work is not stalled. On my own work, I now use only steel, so the nailing gun is obsolete, and I have turned to the screw drill.

With steel studs, use a paper-backed wire fastened with screws, ½-inch (12.7 mm), six-gauge, either Phillips head or Streaker screws. Tie wire can't be fastened to the steel easily, so this eliminates nails and the fastening of tie wire as a backing for regular stucco netting (Fig. 18-8).

Figure 18-8

Lathing with paper-backed wire.

The paper-backed wire has stiffeners in the roll or sheets so that plaster can be applied. Otherwise, the paper would push inward and the application of stucco would be nearly impossible. Paper-backed wire costs more than the regular paper and stucco netting, but the time saved in work might be worth the difference for the inexperienced, even while working over wood (Fig. 18-9).

Plastering

After the application of wire, the job has to be inspected for plastering. On a very small job (such as repair patching), you might

Figure 18-9

(Middle) A parting bead is pictured, applied on expansive walls and especially ceilings to stop continuous cracks in the stucco. Slightly bent corner rite is meant to fit in all corners.

mix cement in a wheelbarrow, but this would not be feasible for an entire home, so you'll need to use a plaster mixer (Fig. 18-10). It really takes an expert to accomplish any of this work, both lathing and plastering. I suggest you hire a licensed subcontractor.

The first coat of cement is applied, scored or scratched with a scratcher. Some inspectors are particular about the scratch job. I used to tell some apprentices that I once had an old rooster that could do a better scratch job than they were doing.

The second coat is applied according to regulations in your area. Some areas require several days' curing time between coats, during which time the walls are to be sprinkled so the cement will harden properly. The inspector might check this coat for hardness, sometimes by rubbing a nickel on the walls. If the stucco scratches the nickel, it has hardened properly.

When applying the second coat, you might sprinkle the first, depending on the weather conditions. Apply a modest area by hand with a trowel, taking the material from a hawk. Straighten out the wall with a darby or rod. When the material hardens somewhat, float

Figure 18-10

Plaster mixer and gun hopper combined. Plenty of cement and the sand pile in front of the mixer. Regular cement is being used, which gets harder than the plastic cement, made for hand-application. Less sprinkling is required for regular cement for a durable, everlasting wall, which could be a problem in a hot, dry climate.

the wall to take out all the bumps, for a better application of the third or final coat.

These days, a job of any size is applied with a plaster gun. The material is mixed and dumped in a hopper, where it is pumped through a hose. Compressed air then blasts the continuous stream of cement against the lathing material—or, in some cases, against the solid wall—of the building. The plasterer manipulates the gun to apply the material evenly (Fig. 18-11). Another worker will scratch the wall with a specially made scratcher. This worker will also use a

Figure 18-11

A man at the gun with his crew may cover several houses in one day. They never shut the gun off. As the gun man climbs onto the scaffolding, a helper/scaffold-mover will manage the application for a few minutes, sprinkling the walls.

brush to clean the material away from windows (generally covered), trim, and corners so the beads show and can then be filled to the proper depth with the second coat.

The third coat is applied with a color, generally a form of cream or yellow. In the United States, a smaller gun is used to apply this finish coat. The material is pumped to the nozzle where a continuous flow of air blows it out in a uniform spray. You can run a trowel over the finish for a texture or leave it with the pimples formed by the gun. As I mentioned before, this is where, in Europe, they now use a heavy gun to apply a substantial finish coat.

In California tracts, where many houses are plastered at one time, a hopper containing perhaps 10 tons (0.9072 tonne) of dry-mixed cement and sand is located on the job (Fig. 18-12). A motor mixes and augers the material to a pump where it is forced to the nozzle. One worker sprays the material over the lath as described before, but several houses are covered in a day with one gun and a four-person crew: One worker operates the plaster machine; another operates the gun; another works the rod, or scratcher, brush, and pail of water for cleanup; and yet another moves scaffolding (Fig. 18-13).

As already stated, stucco as described will form a one-hour firewall. Stucco over cement block or steel studs is even better because if the stucco goes, there is nothing further to feed the fire. I am told that in

Figure 18-12

This tank holds enough premixed dry sand and cement to plaster several houses. A motor at the bottom of the tank mixes the material and augers it into the hopper of a machine that pumps mixture through the hose, where it is blasted onto the exterior walls with forced air, as it oozes from the nozzle of the gun.

Figure 18-13

This man is rodding and floating behind the gun man.

the recent Laguna Beach fire, one house with steel studs and plastered eaves was the only one left standing in a block of wood-frame houses.

In Figs. 18-14 and 18-15, you can see stucco applied over cement block. Both of these buildings are in France, where cement block is

Figure 18-14

Stucco applied over heavy cement block.

Figure 18-15

The finished product, stucco over cement block.

covered with two coats of cement plaster on the exterior wall, the third coat frequently colored.

Stucco roofs

At one time, I stuccoed part of the roof of a house. To be exact, it was in Long Beach, California, where building codes are strict and enforced. The plans called for stucco on the roof to be applied the same as the exterior walls. I nailed down the felt and stucco netting over sheathing. The inspector passed the job, and I put two coats of plastic cement mixed with sand over the roof and floated it out for the finish coat. I applied the finish coat of stucco over the roof the same color as the walls, cream. It looked better as a roof in the neighborhood than did the harsh red tile in evidence everywhere.

This type of roof would not work in a cold climate, with three coats of cement. The finish coat of material would absorb moisture; then, during freezing temperatures, ice would form between the two coats and the expansion would pinch off the color coat. This does not pose too much of a problem on the walls because they are kept relatively dry by the eaves.

In France, as I've mentioned, they have overcome this problem on the walls by applying a heavy finish coat with a gun. With the high cost of tile and shakes, you could turn to the stucco roof as a lasting and cheaper alternative. However, you should do this only with the assistance of the building department. Moisture between the tar paper and the cement might freeze and lift the stucco netting, causing it to become partially unnailed over the years.

Once you've received permission from the building department, you could apply paper-backed wire fastened to screwable rafters (no sheathing). Apply a first coat of stucco, probably with a gun. The second coat would be applied after the first had hardened enough that you could walk on the roof. This coat would be applied evenly and not roded or floated (unless you wanted that effect). Colored rock could be embedded in the second coat of material, but keep in mind that those would, with freezing, loosen from the material and become dislodged, leaving small holes in the stucco.

As a second option, embed the tile into the wet cement of the second coat. This is done in France where a third layer of tile might be applied after 200 years. The tiles used in the past were not glazed (or at least not heavily glazed), and moss grew from these, which is not the case with glazed tile made today.

I am not sure how far north you could apply a tile roof over cement by embedding tile into cement. This again would have to be ironed out with the Department of Building and Safety in northern states.

Take no responsibility for stucco roofs in cold climates. Even in warm climates, stucco roofs need to be experimented with and problems have to be ironed out with the building department in your area. My building experience has been in nearly frost-free regions.

19

Tilt-up

PLANNING is an important factor in any construction project, but especially so in a tilt-up building system. A carefully worked-out schedule of operations is essential to provide the proper sequence of panels ready for the tilt-up process. A casting layout should be drawn up prior to construction to ensure proper use of space and easy access to materials and equipment.

Today, many factories are built with heavy slabs of concrete, poured at the building site over the concrete floor and lifted into position by means of a crane. This is tilt-up construction, so named because the concrete slabs are tilted up to form exterior walls.

This method of forming the walls for a concrete building is much simpler than building forms of plywood and pouring the building solid, a method that has gone out of style, except for an inner support wall or footings below grade level.

There are many advantages of tilting up concrete walls during the process of construction. The building can be lifted into place in one day, while a brick or block crew might be there for weeks. There is little maintenance on a concrete building—no painting, siding, etc.

Of course, concrete-slab walls are fireproof, as well. When these fireproof ideas are introduced more fully into the residential arena, perhaps only a wall or two would be poured to give a delightful contrast to the exterior of the dwelling. Each section would be poured with window openings framed in, which would be set in place later.

Before pouring

The tilt-up walls are framed with 2×6s (5.08×15.24 cm) or thicker, as prescribed by the drawings, perhaps dictated by the height of the building or the number of stories. A webbing of steel is tied in place in two directions, depending on the height of the building and the length of the individual panel (Fig. 19-1), tied together with 18-gauge wire. On a large factory, the slabs may be 18 feet (5.49 m) high and 20 feet (6.096 m) long. It used to be that steel projected from each slab and, after it was set in position, the steel between the slabs and

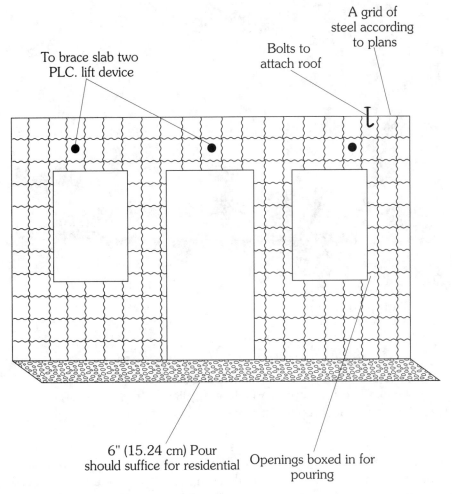

Figure 19-1

A webbing of reinforcing, tied together with 18-gauge tie wire.

at the corners was tied together, perhaps even welded together about a foot (30.48 cm) apart. This was then boxed in and concrete poured-in-place columns were formed.

It is understood that a separating agent is applied to the slab—an oil, a stearic acid, or plaster craft. Different contractors have their favorite means of separating the wall slabs from the floor after they have been poured (Fig. 19-2). The best method might also serve as a

Figure 19-2

Crane lifting a panel into place. Installing panels for housing units would be much simpler.

curing agent for the pour and a bond separation. The first coat to be applied is the curing membrane and should seal the floor surface completely. The final coat should be applied before the wall reinforcing and inserts are in place.

Since the concrete slab is to be used as a casting surface, it should be smoothly troweled. Any imperfections will show up on the new wall. The slab should have a thoroughly compacted subgrade and adequate strength and thickness to support the material trucks or mobile cranes that may be required to operate on it.

Ready-mix is poured into forms nearest to the wall where the slab will be later tilted into place. The pour is leveled off and troweled for a finished product. This type of pour does not generally include windows for industrial, except perhaps along the front wall. With solid pours, when the floor space of the building is not adequate, one slab is sometimes poured over another. In this way, all the walls of

Figure 19-3

A warehouse that will never burn down. A slab will not fall on you if you never get under it.

the building can be tilted in place the same day. Otherwise, a crane has to be moved onto the job site a second time, for another rental fee of perhaps $1,000 (Fig. 19-3).

Curing

After the pour, there is a waiting period while the cement hardens so it does not crumble when lifted into place. This time should be stated on the plans by the engineer so no mistake is made in moving the slabs too soon (Fig. 19-4).

The curing of the panels is very important. The concrete will attain 1,000 psi in one day (453.6 kg per sq. inch)—4,500 psi in seven days (20412 kg. per sq. inch). The slabs should not be lifted until they have attained a minimum of 2,500 psi (11340 kg per. sq. inch)—the reason of waiting a week before lifting the slabs into place.

Figure 19-4

Insert for lifting is embedded in the slab.

Bracing

At the time of pouring, special inserts are incorporated into the slabs along the top to allow the panels to be lifted into place (Fig. 19-4). (The location of the pickup points should be determined by an engineer experienced in this type of work.) The inserts are attached to the rigging and to the crane. The slabs, after the waiting period, are hoisted into place on a foundation and concrete is often poured around the tilt-up panels. Bolts for bracing poles are also incorporated into the panels (Fig. 19-5). A brace will be fastened to the panels before lifting. This is then fastened to a similar device in the floor. The wall is solidly in place, although this is all temporary bracing while the concrete footing hardens around the slabs to hold them in place. In this way, the walls won't fall later in one of nature's various upheavals.

These are good buildings: They are fireproof and will withstand the shocks of nature, especially if they are properly engineered and built accordingly.

Now the roof structure in many instances holds the walls intact. Not quite as good! Holes are drilled into the upper part of the walls or bolts provided in the pour, and heavy beams bolted to the slabs for the roof system. The slabs are held together by the roof system extending from one slab to another.

Later, the lifting devices are removed from the walls and the holes patched, including those in the floor where the braces were in place.

Figure 19-5

A bolt in the floor of the building. The foot of the brace fastened to the bolt.

The attachment at the top of the slab may be used to tie in the roof, or they might have to be removed with a welding torch.

A case study

A tilt-up factory was built in Downey, California, approximately 20,000 square feet (1855.5 sq. meters). The walls had adequate reinforcing steel in two directions. They poured the slabs, walls, on the floor of the proposed building. After the slabs had cured, each 20-foot (6.096 m) section, with built-in devices for lifting, fit against the next when lifted into place. The crane and operator probably cost $1,250 or more per day. It took one long day to set up all the slabs; the crew was paid overtime in order not to have to rent the crane another day.

The sections of wall sat on the slab in a track without sufficient anchorage. The structure should have been better fastened to the foundation, say with a trench and concrete poured around the bottom of the slab as a footing. In this instance, the slabs in the track danced during an earthquake; one end of the building rocked, moved, and jumped, which caused the roof to pull loose from the bolts along the upper part of the wall. This allowed the building to lean over toward the one next door. The roof caved in on one corner but didn't drop entirely. The building was fastened back together, ready for the next quake.

Next door, the cement-block building with sufficient reinforcing steel, adequately fastened to the foundation, faced the quake valiantly, while some of the ten-ton milling machines within slid six inches (15.24 cm) along the floor.

Added reinforcement

In the Mideast where they build fireproof from lack of lumber, many houses and apartments are assembled according to Fig. 19-1. They stand up under the gunfire and cannonading often peculiar to those areas. The buildings might be damaged, but they seldom fall. The webbing of steel remains, so the building can be patched again and again and remains virtually intact.

The walls on such structure have to be fastened down with an earthquake strap, poured with the concrete floor and foundation. The strap is then fastened into the framing of the building so the structure will withstand the first initial shock of a quake or will hold the building from blowing away during a cyclone (see chapter 9). The previously mentioned building was designed to be held together by drilling holes in the slabs at the roof line and attaching 4×10s (10.1×25.4 cm) to the wall for attachment of the roof. Earthquake straps could have been embedded into the foundation, fastened under a reinforcing bar. The strap could then be bolted to the bottom of the slab where holes are drilled for that purpose. This would keep the building from dancing during an major incident where nature freaks out the earth.

For short spans, the walls would be set in place and similar slabs lifted for a roof. Roof slabs would have to be made leak proof at the joints with a special outdoor adhesive compound. Over the joints, the slabs have to be covered with strips of roofing—nails could scarcely penetrate the concrete—hopefully with something that would not blow off in a great wind. There are many ways to cover the joints, tile, brick, strips of poured concrete, etc. Even a pin gun for roofing might fasten down rolled roofing cut into strips.

Insulation

A tilt-up building could scarcely be destroyed, but some type of furring and insulating would have to be added to the inner slab walls so that, in warm climates, the house wouldn't become an oven in the hot summer sun. In the northern states during winter, without furring, frost would accumulate on the inner side of outer walls.

This difficulty could be overcome with four- or six-inch (10.16 or 15.24 cm) steel studs as furring, the insulation inserted to make the buildings suitable for residential use.

Added protection

I feel especially safe from fires when staying in a motel built according to these standards. The walls are tilt-up, properly anchored (Fig. 19-6). There are slabs between floors, extended for a porch or walkway and overhang for the roof of the lower floors.

Anyone hoping to build in a brush-covered, hilly area should look into a house with at least the wall backing up against the combustible jungle built accordingly. Any architect or engineer can design such a unit to make it fireproof. The walls to support such a structure could easily be of tilt-up slabs.

Figure 19-6

Some motels are now built of tilt-up with slabs for floors between two- and three-story units, as well as slabs for the roof.

Tilt-up options

Tilt-up concrete can be provided in a myriad of architectural treatments and shapes (Fig. 19-7). In shopping centers, the customers can be protected from the elements by covered arches or spandrel panels as an arcade or walkway. Long spans are broken up with colored bands in rustication. Some of the features found in commercial construction might not lend themselves to the residential, but there are enough never-ending variations to interest the architect in the innovations of residential tilt-up.

Tilt-up can take on any shape you might desire, rendering your building a work of art that will outlast a dozen fires. It is a crime to see all the homes in the Oakland fire being built with wood a second time when there are so many fireproof products and methods on the market today.

Tilt-up construction has, in a small way, been introduced into residential construction. A few companies do nothing but residential, and, I assure you, those homes will never burn down.

Figure 19-7

A factory like this one will last forever. This method of building is gradually being introduced into the housing industry.

Fireproofing

Various types of tilt-up are completely fireproof, lending themselves to buildings that can become inexpensive housing or million-dollar structures. Concrete can be delivered to the job and poured into any type of forms for fancy designs. As stated, concrete is most extensively used of all building products.

In case of fire in brush-covered areas, the surface of the outer walls of incombustible materials might be slightly damaged from heat or a sudden application of water. If the walls are not damaged structurally, slight flaking can be reconditioned with a steel wire brush. Extensive wall damage can be sandblasted to a solid surface, then covered with a coat of exterior plastering material to bring the house back to normal beauty. In severe cases, Gunite is used, then perhaps a finish coat of plaster or a paint job to finalize the repairs. This is a lot better than rebuilding an entire house of wood.

To further prevent fires, double glass should be used in most construction. If the outer layer is damaged, the inner glass sometimes prevents the flames from entering the house.

In all instances, for the best results, have the building department inspect fire damage and repair the area under their guidance.

Sources

Dayton Superior of Santa Fe Springs, California, has furnished me with pictures of tilt-up for this chapter. They are manufacturers, suppliers, and renters of quality construction accessories needed in forming tilt-up walls. There are innumerable accessories required in the formation, lifting, and bracing of panels—beam hangers, bond breakers, lifting hardware, and several types of bracing so the panels, when secured in place, can be leveled accurately before being embedded permanently in place.

Accessories for lifting tilt-up panels are required mostly in heavy construction, but a minimum might be needed in residential construction where the walls are not over eight or ten feet (2.44 or 3.05 m) in height and are easily shifted into place by less sophisticated machinery than a crane rental at $1,250 per day.

Precast units are another form of tilt-up. It is as simple as taking a cast for duplicating the original pattern, only on a grander scale. The precast units are now used to form four-story buildings. The roof of the cast forms the floor for the next unit. A precast house or half a house will be formed and hauled to the job a week or more after the first pour, where the unit will be lifted into place (Figs. 19-8, 19-9). Or the cast might be hauled to the job and poured on site, then the unit lifted onto the prepared foundation. The form is then hauled to the next job where another pour is made and the unit lifted onto that foundation.

Figure 19-8

Tilt-up slabs on the way to the construction site.

Figure 19-9

A section of the house is lifted in place.

The Portland Cement Association is an organization of cement manufacturers to improve and extend the use of portland cement and concrete through the development, engineering, research, education, and public affairs work. They are located at 5420 Orchard Road, Skokie, Illinois 60077-1083. They can provide invaluable material for those who might want to venture into tilt-up and other uses of cement.

Tierra passive-solar concrete homes

So far all the material on tilt-up construction has been in the form of warehouses, factories, and commercial complexes, but Tierra Homes, I find, is on the forefront in modern concrete technology and has developed a concrete housing unit for the single-family dwelling (Fig. 19-10).

Figure 19-10

At last, a photo of a tilt-up house. Tierra Homes

Fifteen years have gone into perfecting their construction methods to deliver a superior home at a competitive price. As a member of Passive Solar Industries Council, Tierra Homes has state-of-the art computer programs for design and analysis of energy-efficient homes.

Tierra Homes is located at 6898 Lincoln Blvd., Oroville, California, 95966. (The following copyrighted material used with permission.)

The brochure of Tierra Homes states: Concrete is not only an alternative building material, it is the preferred building material. For an industry such as construction, which measures change in decades rather than in minutes or hours, new alternatives to traditional materials and methods have been slow in acceptance. Today, concrete, an old veteran of heavy construction, is finding new life in many light-commercial and residential applications as an alternative to wood framing. It is strong, resistant to fire, termites, and rot, and price-competitive, with an infinite life expectancy.

In the Tierra concrete home, 100% of the exterior walls and foundation are insulated. In traditional wood frame construction, glass fiber insulation is installed between 6-inch (15.24 cm) studs. If all the walls were put together they would represent up to 20% of the total wall area, therefore 20% of the wall is not insulated. Also there is a 25% heat loss through an uninsulated foundation. The interior walls are also concrete for added thermal mass and structural capability.

A concrete home was the only structure to withstand the government atom bomb test "Survival Town" at Yucca Flats Proving Grounds in Nevada during the 1950s.

225

226

Repairs

THERE are many buildings of brick, built in different parts of the world before one thought of using steel as reinforcing to tie them together so they would better resist the forces of nature. Some of these structures are ready to fall into the streets, and many will do so in time. Nearly all of these buildings, even when condemned by the building department can be salvaged and turned into new, safe structures.

Gunite

Gunite might not be considered a viable method of building new walls, but it is excellent for repairs. And preserving or salvaging what is already standing is almost always a better option than hauling the building to our already impressive dumps.

You simply build a grid of reinforcing steel around the entire building—say a framework of ½-inch (12.7 mm) rebars tied together at 12 inches (30.48 cm) OC or closer, depending on the plans, and two inches (5.08 cm) from the exterior wall of the building. Start with either a new foundation or an additional footing adjoining the old, incorporating steel in the pour as you would provide steel for a cement-block or brick building. Then tie 20-foot (6.096 m) pieces of rebar (or as the height of the building dictates) to the upright rebars coming out of the foundation. With a concrete drill, bore holes into the existing brick, block, or sandstone, mostly through the mortar joints and into the interior of the building.

From the inside of the building, punch an L-shaped piece of rebar through the hole and tie the bar to the upright reinforcing outside. This bar might be welded to the grid of outside reinforcing or bent sufficiently to be securely fastened to the new panel of rebars. The bars inside would be embedded into the wall of the old structure so the holes could later be patched in a neat fashion, but only after the outside walls have been Gunited and the reinforcing is solid and in place.

The number of bars joining the old structure to the new grid of reinforcing must be determined by an engineer (who would have drawn up the plans). Some of the L-bars might be shoved only six

228

inches (15.24 cm) into the wall and the foot of these bars fastened to the outer grid. If the brick wall is hollow, shove the L-bars into the wall and fill this hollow area with concrete or grout (which has to be pumped into the void). Be sure the walls are sturdy enough to hold this wet grout and not give way. You might have to pump or pour the void full in stages after the first pour or pumping has set.

When the entire structure has been covered with a grid of steel securely fastened to the building (so no wall will later tumble outward, making a bad situation worse than the previous shaky building), you then need an inspection. This inspection should not be just a cursory glance, to assure that all is going according to plan, but a *real* inspection.

Stretch stainless-steel wires along the building, the entire length of the steel framework, and align the grid of steel to these wires so the new structure will be straight and true. When the building is Gunited, the workers will rod the walls in line with the wires.

You might not be able to form four-, six-, or eight-inch (10.16, 15.24, or 20.32 cm) walls in one pass over the building. If this is the case, you might have to shoot the steel framework only two or three inches (5.08 or 7.62 cm) deep with one pass and later shoot again after the initial pass has set. If you shoot too much material at one time, it might slump, leaving a hollow spot in a wall thought to be solid. After the second or third pass, the walls are roded, leveled, and floated.

In some instances, the finished Gunite walls are painted, but a color coat is better, blown on with a gun on big structures, but often applied by hand on a single residence.

This method of salvaging old buildings needs to be applied to many a shaky brick residence, built before the walls were poured solid—and especially to those without reinforcing steel to hold them together. Hundred-year-old brick houses can easily be Gunited.

Old brownstone might not ever need this method of preservation. If it does, the grid of steel probably won't need to be fastened from within, but the anchors should be driven six inches (15.24 cm) into the drilled holes. This might be sufficient for a one- or two-story building. Depend on the engineer and the plan checker rather than have the walls lean out.

Many old buildings without steel have been renovated in this fashion. A row of classrooms in South Pasadena was Gunited over a grid of steel. I had the contract to apply the finish coat in color over the Gunite. For long walls, earthquake joints were provided, a two-inch (5.08 cm) space where one section of the building could shake separately and not affect the adjoining walls.

I don't believe in tearing any building down, not even old framed houses. (Of course, someday, these will probably burn!) In my small hometown of Spearville, they had built a town hall while I was in high school. This building was recently torn down because it shook. A Gunite job would have tied the building together; maybe only the inner dead air space needs to be pumped full of grout. What a waste.

Sandblasting

Sandblasting is just what the term implies: The blasting onto a building of sharp sand, generally crushed to the point that the sides of each particle have become sharp at the broken area. When a house or any building becomes old or has been painted too many times, it might need to be beautified, brought up to date—The building may only need to be sandblasted to remove old paint on the walls, as well as dirt and grime. Of course, this is not a structural repair but a purely cosmetic one.

Sandblasting is perhaps the best way to restore old beauty. Sand from the river that has been rolled around for a million years is not the best for blasting to remove old paint defacing the structure.

If the building is to have a new coat of finish plaster—Color Coat, as it is called in many instances—then old paint has to be removed. Many old buildings are brought up to date in this way. Stucco is used in the west to cover many jobs and to restore many old buildings. In

230

sandblasting, all the cracks in the wall covering are exposed and the new coat of material will fill these, leaving a new finish coat of colored cement, much better than a coat of paint.

Stucco should be considered sacred. It should not be covered with a coat of paint, which will in time have to be removed. Over time, this could damage the stucco.

In other areas, brick is a surface that requires sandblasting after too many coats of paint. In some cases, the true beauty of the brick may be destroyed by too much paint. In this case, you could recommend a coating of exterior stucco, either blown on or applied by hand. In either case, all the mortar joints will show unless the work is preformed in two applications.

Finding your niche in repairs

Homebuilders, when they understand variations in the building trades, can venture out into areas other than new-builds. There is the least profit in building a house. The most money is made in doing something different, something that scares most bidders away, such as repair jobs. Learn your way around on small jobs, then bid on the bigger jobs. When you venture out to commercial and public jobs, you will be bidding with millionaires and no longer need to work for peanuts.

New York City alone has thousands of obsolete brick buildings in need of renovation. These buildings don't need to be torn down; they can be renovated with Gunite applied over a webbing of reinforcing steel.

Fire prevention

MANY unoccupied buildings are tempting sights to firebugs, people without scruples bent on having a little fun at someone else's expense. For this reason, unoccupied buildings need to be kept in good order so the potential arsonist is not as readily tempted to destroy the structure.

Arsonist might have started some of the fires around Los Angeles in October 1993. None of this has been proven, but it doesn't matter. Prevention comes from thoughtfulness as well as from building fireproof.

The fire in Fig. 21-1 started from the roof, which was the same level as the street in back of the building. Arsonists probably threw something onto the building late at night when there was little traffic to report the fire before it had a good start. The fire that damaged this commercial building probably could have been prevented with a slurry of cement on the roof. The initial fire could not have readily burned through the thin layer of cement. The walls of the building on the street side were of cement block and valiantly withstood the fire. The east wall of stucco over wood did not give way.

Architects, engineers, and homebuilders alike need to step in and suggest that firesafe methods exceed minimum requirements of the building department. Unfortunately, in many instances, trying to persuade one to vary from the standard wood is difficult. A homebuilder could argue that this would cause the loss of too many contracts.

I was, at one time, on the building committee where they proposed to build a $3 million church. I suggested that it be built fireproof, with cement block or brick walls. I was overruled. I suggested fireproof partitions. I was ignored. Only after the billion dollar fires around Los Angeles in the fall of 1993, I was told that the structure would be built of cement block. I had stopped going to the meetings by then, but I am sure the inner partitions will be of lumber. I had told the committee how to build the partitions solid, fireproof, and how to install heavy beams of steel and C-joists to be fireproofed, then covered with 3.4 rib lath (1.54 kg) per square foot (0.092 sq. m). I wrote all the specifications out, even to the size of the screw to be

Figure 21-1

With better building design, fire damage like this could have been prevented.

used to attach the metal lath. I was ignored. So you, too, will have trouble convincing clients that they should build fireproof.

The metal post supports of the roof in Fig. 21-2 should have been wrapped with metal lath and plastered with at least two coats of cement or hardwall. The metal posts in Fig. 21-3 remained upright, but they should have been plastered as stated, for a further fire-safety program. You set a sheet of metal lath upright, and wrap it around the post. The amount of surplus lath isn't worth salvaging, so you double-wrap part of the post. Then tie the metal lath down with

Figure 21-2

A sprinkler system could have prevented this fire damage, but for the price of a sprinkler system, the building could have been built fireproof.

18-gauge wire, at about eight inches (20.32 cm) apart. The post is lathed.

It is difficult to plaster such a curve with a trowel. Dip a slice of 30-pound (13.6080 kg) felt into the mud and pull upward along the post. You will have plastered a third of the post, a distance of one foot (0.3048 m), with the first coat. Use anything as a tool that will readily take the shape of the post, such as a slice of light-gauge metal. In this way, almost anyone can plaster a post. With two coats

Figure 21-3

The beams should have been covered with a shot-on fireproofing.

of plaster or cement, you will prevent the post from slumping in almost any kind of a fire. The city will gladly tell you what they want in the way of prevention, or it should be on your plans.

As a precaution against fire damage, all metal support beams in Fig. 21-3 should have been plastered, and they would not have been severely damaged in the fire. You fasten ⅜-inch (9.5248 mm) reinforcing bars upright every 16 inches (40.64 cm) or every 24 inches (0.6096 m). They would have to be welded to the beam at the top and bottom on both sides. It's not difficult to wire metal lath to the bars. At 24 inches (0.6096 m) OC, 3.4 (1.54 kg) rib lath can be wired in place horizontally. At 16 inches (40.64 cm) OC, a lighter metal can be used.

You can screw rib lath to a ceiling joist. Allow it to drop below the beam, bend the lath onto the surface at the bottom of the beam, and bend upward. Use the same procedure on the other side of the beam. Wire the laps with tie wire. This can be plastered with two coats (in some cases, a third coat might be required). A metal bead might be used on all corners to facilitate plastering.

Generally, some form of covering is required for metal beams. Even if not required, an owner should demand it be noted on his prints. Had the metal beams in Fig. 21-4 been fireproofed, the building could have been saved.

Figure 21-4

Had the metal beams been fireproofed in this structure, it could have been salvaged instead of hauled to the dump.

After lathing and inspection the beams are then plastered with at least two coats of material. In many instances a third coat is applied, smooth, for beauty and especially if the building is to be occupied for food processing. As stated earlier in this book, there is no point in using steel beams if you fail to fireproof them. You can weld 1⅝-inch (26.10 mm) nonbearing sections of screwable metal studs to the beams at 16 inches (40.64 cm) OC. One or two layers of wallboard screwed to the light studs will prevent the beams from sagging in almost any fire.

Figure 21-5

This entire street of frame houses burned in the Oakland fire in 1991.

An entire street of wood-framed houses burned in the Oakland fire in 1991 (Fig. 21-5). Ironically, they have built again with wood.

Figure 21-6 shows the results of a truly devastating fire, Again, much more could have been salvaged had the metal beams and posts been covered with metal lath and plastered. Figure 21-7 shows another fire where block walls remained intact. The more fireproof you build, the more you salvage.

Figure 21-6

More of this building could have been salvaged if the metal beams and posts had been covered with metal lath and plastered.

Figure 21-7

Another fire where block walls remained intact. The more fireproof you build the more you salvage.

The pile of debris in Fig. 21-8 should have been hauled to the dump six years ago; the owner was probably in shock from the fire and it took time a long time for him to recover. At least the iron could have been salvaged.

Fireproof residential construction is finally coming into its own. Don't build before you know how it could be applied to your next building.

Figure 21-8

Building fireproof is the best way to avoid fire damage like this.

Plans and materials

WHEN I decided to familiarize myself with the business of homebuilding again, I first bought a half acre lot for $11,000. I had the weeds scraped up and placed in the center of the lot where there was natural drainage, being careful not to block the natural slope of the land. At $35 per hour, that cost $140—Cheap! (The man knew what he was doing.) I planned to cover the weeds with dirt when the lot was graded prior to building forms for the foundation. It would have cost several hundred dollars to haul the trash to the dump. The land in the desert is depleted of all nutrients from lack of decomposing vegetation. So the weeds when decomposed will be invaluable for growing a garden and flowers.

I paid $650 for plans and another $80 for an environmental report to state how much insulation would be required and the need for double or triple glass, etc. (An R-13 is required for outer walls and R-29 in the ceiling.) I had to have the trusses engineered—another $160. Because I was building inner partitions of rebars and solid plaster, I had to have that engineered—$75—even though channels and rebars have been a recognized method of construction for more than 75 years with various ICBOS by the plastering union as well as building departments. Next, I put down $393 for a plan check. Because the walls were of steel studs, the building department wanted the walls engineered, at the cost of $600. Then I had the lot fenced for $1,650. The engineer will send the calculations directly to the architect and I will, someday, meet him at the building department. This does not include who-knows-how-many hundred dollars for a permit. This has taken five months. The time would have been cut somewhat shorter had I not been busy at writing *Fireproof Homebuilding.*

A lot of this hassle could have been eliminated had I built in wood, but I refuse. Even though charts show the equivalent of steel studs to 2×4s (5.08×10.16 cm) and 2×6s (5.08×15.24 cm) in both #1 and #2 lumber, building departments continue to penalize you for the use of steel. Some day this will be straightened out. I am only trying to set the prime example by using noncombustible materials in the process of building a house.

Material yards

Material for fireproof homebuilding can be found at what is generally termed "material yards." These yards handle nearly everything for building a house except lumber. Lumber yards do not handle steel studs, the only new item on the list of material for building fireproof.

When I mentioned to one dealer that I was writing a book on the uses of steel studs that would replace lumber, he almost went into shock, thinking he would suddenly be put out of business. The changeover is gradual. When I built my house, I had to order the steel from the material yard where I bought my cement, block, and reinforcing. Now I am sure they stock galvanized steel studs and all the accessories mentioned in the book.

There is nothing that says the lumber yard can't also stock steel studs. Many probably do already. Some already handle cement products; brick and block. It all depends on the need in that area.

It's up to you, the homebuilder, to create the need. Then, both material and lumber yards as well as building departments will have to get on the fireproof homebuilding bandwagon.

243

244

Appendix

Materials & tables

Cemco structural steel framing

The information provided in this section is reprinted courtesy of Cemco Structural Steel Framing, 263 Covina Lane, Industry, CA 91744.

C-studs and joists (ecs, cs, xcs & xxc): Manufactured in 12", 10", 8", 6", 4", 3⅝", 3½" and 2½" (304.8 mm, 254.0 mm, 203.2 mm, 152.4 mm, 101.6 mm, 92.07 mm, 88.9 mm, 63.5 mm,) widths from 14, 16, 18, and 20-gauge steel. Flange sizes available are 1⅜", (25.375 mm) and 1⅝" (40.27 with a %16", (14.29 mm) return for extra stiffness. C-studs and joists are available with the web punched or unpunched.

Wide flange studs (WFS): Manufactured in 8", 6", 4", 3⅝" (203.2 mm, 152.4 mm, 101.6 mm, 92.07 mm) widths from 16- and 18-gauge steel, with 1⅜" (34.92 mm) flange.

Standard flange studs (SFS): Manufactured in 8", 6", 4", 3⅝" and 2½" (203.2 mm, 152.4 mm, 101.6 mm, 92.7 mm, 63.5 mm) widths, from 16- and 18-gauge steel, with a 1" (25.4 mm) flange.

Deep leg track (DLT, SDLT, SMLT): Manufactured in 12", 10", 8", 6", 4", 3⅝", 3½" and 2½" (304.8 mm, 254.0 mm, 203.2 mm, 152.4 mm, 101.6 mm, 92.07 mm, 88.9 mm, 63.5 mm) widths from 16-gauge steel. 8", 6", 4", 3⅝" 3½" and 2½" (203.2 mm, 152.4 mm, 101.6 mm, 92.07 mm, 88.9 mm, 63.5 mm,) widths from 18-gauge steel web unpunched.

Standard leg track (SLT): Manufactured in 10", 8", 6", 4", 3⅝", 3½" and 2½", (254.0 mm, 203.2 mm, 152.4 mm, 101.6 mm, 92.07 mm, 88.9 mm, 63.5 mm), widths from 16-gauge steel. 8", 6", 4", 3⅝", 3½", and 2½" (203.2 mm, 152.4 mm, 101.6 mm, 92.07 mm, 88.9 mm, 63.5 mm) from 18-gauge steel. Web is unpunched.

Materials

Sections (14- and 16-gauge) are fabricated from steel conforming to ASTM-A570 grade 50, with a minimum yield point of 50,000 psi (22,680 kg). Sections (18- and 20-gauge) are fabricated from steel conforming to ASTM-A611 Grade C, with a minimum yield point of 33,000 psi (14,968.8 kilograms). All sections are painted with rust-inhibited paint or also available with galvanized finish. When galvanized finish is specified, material conforming to ASTM-A446 will be furnished.

Utility

Cemco structural steel framing components are ideally suited for load-bearing walls, roof joists, and floor joists and are compatible with all types of building materials and techniques of construction.

Storage of materials

Materials should be covered or protected from rain, snow, or other corrosive elements that may damage the materials. Special attention should be noted on exposed-job-site condition. Material should be stored on a flat plane. Do not overstack the pallets or you could damage the material.

Section properties

The section properties set forth in Table A-1 for Cemco structural steel framing are based upon the specifications for design of cold-formed steel structural members of the American Iron and Steel Institute (U.B.C. standard no. 27-9).

The section properties are used as a basis for deriving the Cemco load and span tables which will provide most design needs. For special conditions the section property tables may be used by the designer to develop allowable loads and spans.

Series "ECS"—Punched "C" studs and joists, 1⅜" flange

Table A-1

SECTION DESIGNATION	DIMENSIONS (IN.)					GA.	WEIGHT
	DEPTH	FLANGE	PUNCHOUT	RETURN	THICKNESS		
	—	—	L x W	—	—	—	LB./FT.
250ECS14P	2-1/2	1-3/8	L = 2.000 W= 0.750		0.0747	14	1.400
250ECS16P					0.0598	16	1.136
250ECS18P					0.0478	18	0.918
250ECS20P					0.0359	20	0.697
350ECS14P	3-1/2	1-3/8	L = 3.250 W= 1.500		0.0747	14	1.574
350ECS16P					0.0598	16	1.275
350ECS18P					0.0478	18	1.029
350ECS20P					0.0359	20	0.780
362ECS14P	3-5/8	1-3/8	L = 3.250 W= 1.500		0.0747	14	1.607
362ECS16P					0.0598	16	1.301
362ECS18P					0.0478	18	1.050
362ECS20P				9/16	0.0359	20	0.796
400ECS14P	4	1-3/8	L = 3.250 W= 1.500		0.0747	14	1.704
400ECS16P					0.0598	16	1.379
400ECS18P					0.0478	18	1.112
400ECS20P					0.0359	20	0.843
600ECS14P	6	1-3/8	L = 3.250 W= 1.500		0.0747	14	2.074
600ECS16P					0.0598	16	1.674
600ECS18P					0.0478	18	1.348
600ECS20P					0.0359	20	1.019
800ECS14P	8	1-3/8	L = 4.250 W= 2.500		0.0747	14	2.581
800ECS16P					0.0598	16	2.081
800ECS18P					0.0478	18	1.673

247

Cemco also provides load tables to establish allowable spans for Cemco structural and partitions steel framing members, incorporating variable spacings based upon the most frequently used combinations of dead, live, wind, and snow loads. The data covers a wide variety of framing elements including rafters, ceiling joists or runners, floor joists,

Table A-2   **Series ECS Punched C-studs and joists, 1⅜" flange (34.92 mm) SI Conversion**

Designation	Depth	Flange	Punchout	Return	Thickness	ga.	kg.
250ECS14P	2½	34.92	L=50.8	.75	1.98	14	.65
250ECS16P	"	"	W=19.5	"	1.19	16	1.136
250ECS18	"	"	"	"	1.18	18	0.918
250ECS20	"	"	"	"	0.90	20	0.697
350ECS14P	3½"	"	L=3.250	"	1.98	14	1.574
350ECS16P	"	"	W=1.500	"	1.19	16	1.1275
350ECS18P	"	"	"	"	1.18	18	1.029
350ECS20P	"	"	"	"	0.90	20	0.780
362ECS14P	3⅝"	"	L=3.250	"	1.90	14	1.6072
362ECS16	"	"	W=1.500	"	1.19	16	1.301
362ECS18P	"	"	"	"	1.18	18	1.050
362ECS20P	"	"	"	"	0.90	20	0.796
400ECS14P	4	"	"	"	1.90	14	1.704
400ECS16P	"	"	"	"	1.19	16	1.379
400ECS18P	"	"	"	"	1.18	18	1.112
400ECS20P	"	"	"	"	0.90	20	0.843
600ECS14	6	"	"	"	1.90	14	2.074
600ECS16P	"	"	"	"	1.19	16	1.674
600ECS18P	"	"	"	"	1.18	18	1.348
600ECS20P	"	"	"	"	0.90	20	1.019
800ECS14P	8	"	L=250	"	1.90	14	2.581
800ECS16	"	"	W=2.50	"	1.19	16	2.081
800ECS18	"	"	"	"	1.18	18	1.673

Effective section properties. Torsion/flectural properties not included.

Series "XXC"—Punched "C" joist spans (2½" flanges)   Table A-3

DESIGN LOADS (PSF)	JOIST SPACING (IN)	600 XXC			800 XXC			1000 XXC		1200 XXC	
		14 GA.	16 GA.	18 GA.	14 GA.	16 GA.	18 GA.	14 GA.	16 GA.	14 GA.	16 GA.
Ixx DEFLECTIONS (IN.⁴)		4.945	3.974	3.206	9.424	7.519	6.052	15.582	12.354	23.466	18.498
RESIST MOMENT (IN LBS.)		41,570	31,734	17,816	57,234	43,404	24,390	72,807	54,986	88,862	66,634
10# D.L.	12	22'1"	20'6"	19'1"	27'4"	25'5"	23'3"	32'4"	29'11"	37'1"	34'3"
+20# L.L.	16	20'1"	18'8"	17'2"	24'10"	23'1"	20'1"	29'5"	27'3"	33'9"	31'2"
=30# T.L.	24	17'6"	16'3"	14'0"	21'9"	20'2"	16'5"	25'8"	23'9"	29'5"	27'2"
10# D.L.	12	16'3"	15'1"	14'0"	20'2"	18'8"	16'5"	23'10"	22'1"	27'4"	25'3"
+50# L.L.	16	14'9"	13'9"	12'2"	18'4"	17'0"	14'3"	21'8"	20'0"	24'10"	22'11"
=60# T.L.	24	12'11"	12'0"	9'11"	16'0"	14'10"	11'7"	18'11"	17'5"	21'8"	19'2"
10# D.L.	12	12'11"	12'0"	10'4"	16'0"	14'10"	12'1"	18'11"	17'6"	21'8"	20'0"
+100# L.L.	16	11'8"	10'11"	8'11"	14'6"	13'6"	10'5"	17'2"	15'9"	19'8"	15'11"
=110# T.L.	24	10'3"	9'6"	7'4"	12'8"	11'5"	6'11"	14'10"	12'1"	16'4"	10'7"
20# D.L.	12	20'1"	18'8"	17'2"	24'10"	23'1"	20'1"	29'5"	27'3"	33'9"	31'2"
+20# L.L.	16	18'2"	16'11"	14'11"	22'7"	20'11"	17'5"	26'9"	24'9"	30'8"	28'3"
=40# T.L.	24	15'11"	14'9"	12'2"	19'9"	18'3"	14'3"	23'4"	21'4"	26'9"	23'6"
20# D.L.	12	16'3"	15'1"	13'0"	20'2"	18'8"	15'2"	23'10"	22'1"	27'4"	25'2"
+50# L.L.	16	14'9"	13'9"	11'3"	18'4"	17'0"	13'2"	21'8"	19'9"	24'10"	21'9"
=70# T.L.	24	12'11"	12'0"	9'2"	16'0"	14'4"	10'9"	18'7"	16'2"	20'6"	16'9"
20# D.L.	12	12'11"	12'0"	9'11"	16'0"	14'10"	11'7"	18'11"	17'5"	21'8"	19'2"
+100# L.L.	16	11'8"	10'11"	8'7"	14'6"	13'5"	9'7"	17'2"	15'1"	19'2"	14'7"
=120# T.L.	24	10'3"	9'4"	7'0"	12'7"	10'11"	6'4"	14'2"	11'1"	15'8"	9'9"
30# D.L.	12	18'7"	17'4"	15'4"	23'1"	21'5"	18'0"	27'4"	25'3"	31'4"	28'11"
+20# L.L.	16	16'11"	15'9"	13'4"	21'0"	19'5"	15'7"	24'10"	22'11"	28'5"	25'9"
=50# T.L.	24	14'9"	13'9"	10'10"	18'4"	17'0"	12'9"	21'8"	19'1"	24'4"	21'0"
30# D.L.	12	15'11"	14'9"	12'2"	19'9"	18'3"	14'3"	23'4"	21'4"	26'9"	23'6"
+50# L.L.	16	14'5"	13'5"	10'6"	17'11"	16'5"	12'4"	21'2"	18'6"	23'6"	20'4"
=80# T.L.	24	12'7"	11'5"	8'7"	15'5"	13'5"	9'7"	17'5"	15'1"	19'2"	14'7"
30# D.L.	12	12'11"	12'0"	9'6"	16'0"	14'10"	11'2"	18'11"	16'9"	21'4"	18'0"
+100# L.L.	16	11'8"	10'11"	8'3"	14'6"	12'11"	8'10"	16'8"	14'6"	18'5"	13'6"
=130# T.L.	24	10'3"	9'0"	6'9"	12'1"	10'6"	5'10"	13'7"	10'2"	15'1"	9'0"

nonbearing interior and exterior wall studs subjected to combined vertical and horizontal loads. The allowable loads and spans set forth in these tables are based upon accepted deflection, bending shear, and axial load design criteria set forth in the Uniform Building Code and other nationally recognized standards. Floor joists deflection criteria are based upon members having a maximum deflection of L/360 for live load and L/240 for combined live load and dead loads.

249

Table A-4   **Series XXC Punched C-Joists spans 2½" (63.5 mm) flanges SI equivalent**

Design load (psf)	Joist spacing (m)	600 XXC			800 XXC		
		14 ga.	**16 ga.**	**18 ga.**	**14 ga.**	**16 ga.**	**18 ga.**
Ixx deflect. (mm.)		125.8	86.12	81.36	229.2	190.5	153.90
Resist mov. in lbs.		41,570	31,734	17,816	57,234	43,404	24,390
Kilograms		18,856	14,194	8,083	25,961	19,689	11,063
10# d.l.	3.048	6.73	5.82	6.05	8.53	7.73	7.08
20# l.l.	4.064	6.12	5.23	5.71	7.47	7.04	6.12
30# t.l.	6.096	5.35	5.05	4.27	6.78	6.15	5.1
10# d.l.	3.048	5.054	4.83	4.27	6.15	5.69	5.1
50# l.l.	4.064	4.50	4.19	3.70	5.89	5.20	5.03
60# t.l.	6.096	3.93	3.66	3.02	4.05	4.22	3.53
10# d.l.	3.048	3.93	3.66	3.56	4.98	4.5	3.68
100# l.l.	4.064	3.55	3.33	2.72	4.40	4.11	3.2
110# t.l.	6.096	3.13	2.90	2.24	3.87	3.48	2.1
20# d.l.	3.048	6.12	5.69	5.23	7.87	7.04	6.15
20# l.l.	4.064	5.54	5.26	4.55	6.88	6.5	5.33
40# t.l.	6.096	4.85	4.50	3.71	6.02	5.56	5.35
20# d.l.	3.048	5.05	4.60	3.96	6.15	5.69	4.62
50# l.l.	4.064	4.50	4.19	3.43	5.69	5.20	4.01
70# t.l.	6.096	3.93	3.66	2.79	4.98	4.78	3.28
20# d.l.	3.048	3.94	3.66	3.25	4.98	4.52	3.53
100# l.l.	4.064	3.56	3.33	2.61	4.42	4.09	2.92
120# t.l.	6.096	3.12	2.84	2.13	3.83	3.32	1.93
30# d.l.	3.048	5.66	5.30	4.67	7.04	6.53	5.69
20# l.l.	4.064	5.26	1.80	4.06	6.40	5.91	4.75
50# t.l.	6.096	4.54	4.19	3.30	5.59	5.20	3.88
30# d.l.	3.048	4.84	4.54	3.71	6.02	6.25	4.35
50# l.l.	4.064	4.39	4.09	3.20	5.48	5.11	3.76
80# t.l.	6.096	3.84	3.48	1.90	4.70	4.09	2.92
30# d.l.	3.048	3.94	3.66	2.90	4.90	4.50	3.40
100# l.l.	4.064	3.55	2.51	3.48	4.42	3.94	2.97
130# t.l.	6.096	3.12	2.74	3.12	4.17	3.20	1.78

		1000 XXC		1200 XXC	
		14 ga.	**16 ga.**	**14 ga.**	**16 ga.**
Ixx deflect. (mm.)		15.582	12.354	23.466	18.498
Resist mov. in lbs.		72,807	54,986	88,862	66,634
Kilograms					
Design load (psf)	Joist spacing (m)				
10# d.l.	3.048	10.85	9.10	11.27	10.44
20# l.l.	4.064	8.95	8.62	10.29	9.50
30# t.l.	6.096	7.82	7.24	8.95	8.59
10# d.l.	3.048	7.26	6.73	8.64	7.69
50# l.l.	4.064	6.60	6.10	7.47	6.98
60# t.l.	6.096	5.77	5.31	6.60	5.84
10# d.l.	3.048	5.76	5.33	6.60	6.10
100# l.l.	4.064	5.23	4.80	6.04	4.85
110# t.l.	6.096	4.52	3.68	5.08	3.23
20# d.l.	3.048	8.96	8.62	10.29	9.50
20# l.l.	4.064	8.53	7.44	9.35	8.61
40# t.l.	6.096	7.11	6.50	8.05	7.16
20# d.l.	3.048	7.47	6.73	8.64	8.12
50# l.l.	4.064	6.60	6.70	7.47	6.63
70# t.l.	6.096	5.64	5.03	6.25	5.21
20# d.l.	3.048	5.77	5.31	6.60	5.84
100# l.l.	4.064	5.23	4.62	5.84	4.44
120# t.l.	6.096	4.30	3.38	4.77	2.97
30# d.l.	3.048	8.64	7.69	9.55	8.81
20# l.l.	4.064	7.47	6.98	8.66	7.84
50# t.l.	6.096	6.60	5.82	7.32	6.40
30# d.l.	3.048	7.11	6.50	8.05	7.16
50# l.l.	4.064	6.45	5.64	7.16	6.20
80# t.l.	6.096	5.30	4.60	5.84	4.44
30# d.l.	3.048	5.77	5.20	6.50	5.79
100# l.l.	4.064	5.18	4.42	5.61	4.11
130# t.l.	6.096	4.14	3.10	4.60	2.74

Recognition

See ICBO report No. 3403p for allowable values and/or conditions of use concerning material presented in this document. It is subject to reexamination, revision and possible cancellation.

Structural steel framing

Section properties set forth in Table A-1 are based on the provisions of the U.B.C. and the design criteria of the American Iron and Steel Institute. The material thickness shown are for uncoated material. The allowable compressive stresses are based upon yield of 50,000 lbs./sq. in. (22,580 kilograms) for No. 14 and 16 gauge and 33,000 lbs./sq. in. (14,968.8 kilograms) for No. 18 and 20 gauge.

From wood to metal

There are so many charts of studs and joists that all could not be enumerated here, but I have included just enough to let you know what is available for the homebuilder who might want to gradually change from wood to metal (Table A-5). The charts are invaluable for the engineer so he will be able to know what a 20-gauge stud will support and what a joist will span. An engineer can obtain a complete book of tables from the manufacturer and be listed as one capable of designing or engineering that manufacturer's products. In many cases, an engineer might need only a few pages, which the material distributor will be able to furnish. The building department might request that the contractor furnish a list of the materials he will be using and the name of the product. That should be listed on the prints.

The table gives you the equivalent of a 2×4 in steel. The building departments need to begin recognizing this as a viable information.

Western metal lath preliminary wood framing to metal stud conversion tables

Vertical load-bearing wall studs

2×4 df #1 (50.8×101.6 mm)	3½" (88.8 mm) wcs 20 ga or 18 ga
2×4 df #2 (50.8×101.5 mm)	3½" (88.8 mm) wcs 20 ga or 18 ga
2×6 df #1 (50.8×153.0 mm)	6" (153.0 mm) wcs 20 ga or 18 ga
2×6 df #2 (50.8×153.0 mm)	6" (153.0 mm) wcs 20 ga or 18 ga

Joists and rafters

2×6 (50.8×153.0 mm) df #1	6" (153.0 mm) wcs 18 ga
2×6 (50.8×153.0 mm) df #2	6" (152.0 mm) wcs 20 ga
2×8 (50.8×203.0 mm) df #1	8" (203.0 mm) cj 18 ga
2×8 (50.8×203.0 mm) df #2	8" (203.0 mm) wcs 18 ga
2×10 (50.8×253.8 mm) df #1	10" (253.8 mm)

Fasteners for steel

I have placed throughout the book suggested tools and fasteners for the assembly of the steel-stud house, but here are a few tables of recommendations by a company supplying tools and accessories.

Table A-6

Self drilling screws

	Part no.	Size	Point no.	Cart.	Wt lbs
Bugle Phillips	11	6×1	2	10m	32
	11a	6×1⅛	2	10m	35
	12	6×1¼	2	8m	31
	13	6×1⅝	2	5m	25
	14	6×1⅞	2	4m	22
	16	8×2⅜	2	2.5m	20
	17	8×2⅝	2	1.6m	16
	18	8×3	2	1.4m	18
Pan framing	23	7×7⁷⁄₁₆	2	10m	27
Pancake "	35e	10×⅝	3	7.5m	38
K-lath	30	8×½	2	10m	27
	31	8×¾	2	10m	47
	32	8×1	2	10m	47
	33	8×1¼	2	6m	40
	33A	8×1⅝	2	4m	34
	33AC	8×1⅞	2	3m	32
Pan Phillips	40	6×⅜	2	20m	37
	24	8×½	2	10m	33
	36	10×⅝	2	10m	41
	37	8×1	2	8m	42
	34	10×½	2	8.5m	39
	35	10×⅝	3	7.5m	36
	38	10×¾	3	6.5m	37
	39	10×1	3	5m	44
Hex washer	25	8×½	2	10m	37
Hex ¼ head	25s	8×½	2	10m	37
	21	8×⅝	2	10m	40
	26	8×¾	2	10m	46
	27	8×1	2	8m	44

	Part no.	Size	Point no.	Cart.	Wt lbs
⁵⁄₁₆ Hex	45	10×½	2	8.5m	48
	22	10×⅝	3	7.5m	47
	28	10×¾	3	6.5m	45
	29	10×2	3	5m	42
	65	10×1½	3	4m	40
	70	12×¾	3	5m	45
	71	12×1	3	3m	46
	72	12×1½	3	2.5m	36
	73	12×2	3	2m	35
⅜ Hex	80	14×¾	3	4m	53
	81	14×1	3	3m	41
	82	14×1¼	3	2.5m	40
	83	14×1½	3	2m	42
	84	14×2	3	1m	25
⁵⁄₁₆	150	12×⅞	3	5m	46
	155	12×1¼	3	1m	38
Wafer	90	10×¾	3	7.5m	45¼ Ply
Phillips	91	10×1	3	5m	34⅜ Ply
	92	10×1¼	3	4m	36½ Ply
	93	10×1½	3	3m	31½ Ply
	94	10×1⁷⁄₁₆	3	3.5m	33⅝ or ¾ Ply to .175
Metal flat Phillips	100	10×1⅝	3	3m	31
	101	10×2	4	1.4m	26
	103	12×2	4	1.5m	32
	105	¼×2¾	4	2m	32
Trim head	60	6×1⅝	3	5m	25
	61	6×2½	3	5m	25 metal

Appendix

Table A-7

Self drilling screws in mm

	Part no.	Size	Point no.	Cart.m	Kilo.
Bugle Phillips	11	6×25.4	2	10m	14.515
	11a	6×28.58	2	8m	14.87
	12	6×31.75	2	8m	14.0616
	13	6×41.27	2	5m	11.3250
	14	6×47.62	2	4m	9.979
	16	8×60.32	2	2.5m	9.072
	17	8×66.67	2	1.6m	7.2576
	18	8×76.2	2	1.4m	8.1540
Pan framing	23	7×11.11	2	10m	12.25
Pancake "	35e	10×15.87	3	7.5m	17.24
k-Lath	30	8×12.67	2	10m	12.25
	31	8×19.4	2	10m	21.31
	32	8×25.4	2	10m	21.31
	33	8×31.74	2	6m	18.14
	33a	8×41.27	2	4m	15.42
	33ac	8×47.62	2	3m	14.51
Pan Phillips	40	6×9.52	2	20m	16.78
	24	8×9.53	2	10m	14.96
	36	8×19.5	2	10m	18.60
	37	8×25.3	2	8m	19.05
	34	10×12.7	2	8.5m	17.69
	35	10×15.87	3	7.5m	16.33
	38	10×19.5	3	6.5m	16.78
	39	10×24.4	3	5m	19.96
Hex washer	25	8×12.7	2	10m	16.78
Head 6.35 Hex	25s	8×12.7	2	10m	16.78
	21	8×15.87	2	10m	18.14
	26	8×19.5	2	10m	20.87
	27	8×25.4	2	8m	19.96
7.9 Hex	45	10×12.7	2	8.5m	21.77
	22	10×15.87	2	7.5m	21.32

	Part no.	Size	Point no.	Cart.m	Kilo.
7.9 Hex	28		2	6.5m	20.41
	29	10×25.4	3	5m	19.05
	65	10×38	3	4m	18.14
	70	12×19.5	3	5m	20.41
	71	12×25.4	3	3m	20.87
	72	12×38	3	2.5m	16.33
	73	12×50.8	3	2m	14.88
9.52 Hex	80	14×19.5	3	4m	24.04
	81	14×25.4	3	3m	18.60
	82	14×31.75	3	2.5m	18.14
	83	14×38.1	3	2m	19.05
	84	14×50.8	3	1m	11.34
7.93 Hex	150	12×22.22	3	5m	20.86
	155	12×31.75	3	1m	17.24
Wafer	90	10×19.5	3	7.5m	20.41 or 6.35mm ply
Phillips	91	10×25.4	3	5m	15.42 or 9.52 ply
	92	10×31.75	3	4m	16.33 or 12.5 ply
	93	10×38.1	3	3m	14.06 or 12.5 " ply
	94	10×42.86	3	3.5m	14.97 or 15.87 or 19.5 ply. to 4.36
Metal flat Phillips	100	10×41.27	3	3m	14.06
	101	12×50.8	4	1.5m	11.79
	103	12×62.15	4	1.5m	14.52
	105	6.35×69.14	4	1m	14.52
Trim head	60	6×41.27	3	5m	11.34 wd.
	61	6×57.14	3	3m	11.34 metal

Table A-8

Screws

Ctn qty		Unit price
10m	#8×½ (12.7 mm) Wafer head tek	10.95m
7.5m	#10×⅝ (0.625 mm) Lo-profile tek	16.95m
6.5/8m	#10×¾ (19.5 mm) Hex head tek	16.95m
3m/4m	#10×1½ (38.1 mm) Ply tek	29.95m

Table A-9

Accessories

Qty		Unit price
100	#2 Bit tips	.35ea
12	Mag bit tip holder	4.50ea
12	⁵⁄₁₆(7.94 mm) Mag Hex shafts(short 1⅝)	3.50ea
12	⁵⁄₁₆(7.94 mm) Mag Hex shaft (long 2⁹⁄₁₆)	3.50ea
10	Chop saw cut-off wheels	5.95ea

Table A-10

Power tools

Qty		Unit price
1	Black and Decker 2038 screw gun w/50' cord (5. amp)	119.00ea
1	Black and Decker 2054 versaclutch tek driver	185.00ea
1	Black and Decker 3935 13 amp chopsaw	229.00ea
1	Black and Decker 2731 15 amp chopsaw	270.00ea
1	Kett E-440 14 ga power shears	170.00ea

Hand tools

Qty			Unit price
1	Visegrip	6r C clamps	8.95ea
1	Visegrip	11r C clamps	14.95ea
1	Prosnip	Hand snips	10.95ea
1	Prosnip	Off set hand snips	13.95ea
1	Empire	Torpedo levels	7.95ea
1	Empire	4 ft.(1.2192 mtr.) mag. levels	39.00ea
1	Empire	6 ft.(1.8288 mtr.) mag. levels	75.00ea
1	Empire	Rafter lay-out square	6.95ea
1	Greenlee	Stud punch	159.00ea

Adapted from: Aabatix: Aabatix Environmental Corporation, 12316 Bell Ranch Dr., Santa Fe Springs, CA 90670

Guidelines for approximating material takeoff for steel framing

Exterior wall	L/F × 4 or 6	= track.
Interior wall	L/F × 3 or ?	= track.
Total wall	L/F × 1	= studs.
Headers	R/O + ^" (127.4 mm)	= header material.
Joists	Piece count +2 at bathroom (s) for plumbing.	
Sheer plates	Panel × 4 = plates.	
Sheer straps	Panel × 2 = straps.	
Strap length	A1 (wall height) + (panel length + C1 (strap length >)	
Roof trusses	Truss calcs needed for material take-off.	
Fascia material	L/F × 1.05 = facia material.	
Ridge cap	L/F × 1.05 = ridge material.	

All material should come in lengths with 2-inch increments starting at eight feet up to 40 feet. Cut to length only if quantity equals 40 or more. Most material yards stock standard studs and track, so you can buy and haul your own as needed. Material tolerances = +0 or −¼.

Index

Illustrations are in **boldface**.

About the Author

A licensed contractor, Leo Du Lac has worked in construction most of his life. His work in restoring fire-damaged buildings prompted his interest in fireproof building and fire prevention from a residential construction viewpoint, and this book is the result of over 15 years of research. Using the European fireproof methods as a base, Mr. Du Lac has learned from the mistakes and successes of others to put together this informative book for homebuilders. He has previously written *Gardening in the Dry Lands* (1991, Desert Books) as well as numerous columns and articles for construction trade publications.

Other Bestsellers of Related Interest

**The Portland Cement Association's Guide to Concrete
Homebuilding Systems**
—Pieter VanderWerf/W. Keith Munsell
The first comprehensive sourcebook available on concrete-based
homebuilding systems, this guide was written by two members of the
Portland Cement Association—one of the major contributors to this
year's New American Home, the "idea" house built specifically for
the 1994 National Association of Home Builders Show. Featuring a
color section of photographs of the New American Home.
0-07-067020-X $42.95 Hardcover

Residential Steel Framing Handbook
—Robert Scharff and The Editors of Walls & Ceilings *Magazine*
This book provides in-depth coverage of steel framing, discussing the
advantages and thoroughly explaining the techniques. Valuable
features include reference charts that outline standards and materials
costs, information on the newest materials and tools, and the latest
details on the code-exceeding aspects of steel framing.
0-07-057231-3 $50.00 Hardcover